Springer Actuarial

This is a series on actuarial topics in a broad and interdisciplinary sense, aimed at students, academics and practitioners in the fields of insurance and finance.

Springer Actuarial informs timely on theoretical and practical aspects of topics like risk management, internal models, solvency, asset-liability management, market-consistent valuation, the actuarial control cycle, insurance and financial mathematics, and other related interdisciplinary areas.

The series aims to serve as a primary scientific reference for education, research, development and model validation.

The type of material considered for publication includes lecture notes, monographs and textbooks. All submissions will be peer-reviewed.

Runhuan Feng

Decentralized Insurance

Technical Foundation of Business Models

Springer

Runhuan Feng
Department of Mathematics
University of Illinois
Urbana, IL, USA

ISSN 2523-3262 ISSN 2523-3270 (electronic)
Springer Actuarial
ISBN 978-3-031-29558-4 ISBN 978-3-031-29559-1 (eBook)
https://doi.org/10.1007/978-3-031-29559-1

Mathematics Subject Classification: 91G20, 91G80, 91B30

This Springer imprint is published by the registered company Springer Nature Switzerland AG
The registered company address is: Gewerbestrasse 11, 6330 Cham, Switzerland

You never change things by fighting the existing reality. To change something, build a new model that makes the existing model obsolete.
—Buckminster Fuller

Foreword by Marco Mirabella

Professor Runhuan Feng was one of the first people I reached out to, during the first days of my company, Ensuro. His research on peer-to-peer risk sharing, mutual aid, and decentralized insurance has given me and my team several innovative and interesting perspectives on the way we were looking at the problem of decentralized risk pools. Since then, Prof. Runhuan Feng has become a friend and a learning partner. His strong quantitative background, extensive academic experience, and innate curiosity made him one of the people I call most for advice.

This book gives a holistic view of the world of decentralized insurance. It touches on the history and the reason why decentralized insurance exists. It then analyzes the structures and the laws that it relies upon.

I hope this book will be a catalyst for the new generation of talents, to take part in the decentralized insurance ecosystem and foster ideas and innovations that will be impactful for the lives of many.

Taipei, Taiwan
October 2022

Marco Mirabella

Marco Mirabella is the Chief Executive Officer of Ensuro, a decentralized reinsurance company licensed by the Bermuda Monetary Authority. Prior to Ensuro, Marco was the Chief Strategy Officer at BigGo, the biggest price comparison website in South East Asia. Marco is also the Founder of Cartesi, a layer2 blockchain solution. He started his career in the venture capital space, working for SOSV a venture capital fund with US$1 billion asset under management.

Foreword by Michel Denuit and Christian-Yann Robert

We have known our colleague Prof. Runhuan Feng for his great contributions to the actuarial mathematics of decentralized insurance, rooted in the context of the transition to sharing economy. He has been among the first actuaries to draw the attention of the profession to these new insurance business models, including peer-to-peer insurance in the West, mutual aid platforms in Asia, and takaful in the Middle East. Professor Feng has published several inspiring works that benefitted our own research on the topic. It is therefore our great pleasure to write this foreword for his new book, *Decentralized Insurance: Technical Foundation of Business Models*, which offers the first comprehensive study of economic and actuarial models supporting decentralized insurance.

Feeling to belong to a community has become nowadays important for many individuals. In that respect, commercial insurance does not seem to be very appealing. This may be attributed to the progressive demutualization of the insurance sector, which has broken the link with the ancestral compensation mechanism consisting in using the contributions of the many to balance the misfortunes of the few. In order to remedy this situation, new offers started to develop in line with the sharing economy, with community-based online platforms acting as insurance marketplaces, sometimes surrounded by regulatory uncertainty.

This book considers new insurance business models, ranging from pure Decentralized Insurance (DeIn) where no commercial insurer is needed anymore, to new generations of participating insurance policies where part of the risk is shared within a community and higher losses, exceeding the community's risk-bearing capacity, are still covered by an insurance company. This innovative approach is expected to mitigate the conflict inherent to commercial insurance (where insurers keep the part of premiums that is not paid out for claims), to reduce fraudulent behavior, and to lower prices. It can also help to render insurance more transparent, making explicit the split into taxes, intermediary's commission, claim costs (including settlement expenses, on an incurred basis), running expenses, shareholders' profit, and surplus distributed (cashback) or allocated (give-back) to a common project. This would help to demystify insurance (up to outputs of actuarial models calculating incurred losses or allocating capital) and revives the feeling to belong to a community, helping its unlucky members (like in old times). The development of most of these new business models will have to rely on technological innovations such as Decentralized Finance (DeFi) and will be built on the ability to execute contracts between parties with little to no human intervention. A high degree of automation is expected to make policies affordable (even for very low sums insured) and to rapidly payout claims.

This book is both timely and needed. Even if mathematical foundations date back to the 1960s with seminal contributions after Karl Borch, actuaries specifically considered DeIn quite lately, decades after the first market innovations were launched with varying degrees of success. Professor Feng is well-equipped to write the first book proposing actuarial models for emerging DeIn markets. He is a Fellow of the Society of Actuaries and a Chartered Enterprise Risk Analyst. He currently serves on the SOA's Research Executive Committee and was previously the Chair of the SOA Education and Research Section Council. He received grants from the Casualty Actuarial Society to develop research on DeIn. He co-founded the Illinois Risk Lab and serves as an independent consultant to many corporations and startups. His numerous contacts with the industry and his academic background guarantee that he can identify the topics that are relevant to actuaries.

There have been many actuarial textbooks published over the past 10 years and one might ask whether there is a need for another one. The pioneering work conducted by Prof. Feng provides a convincing analysis of the actuarial models supporting DeIn, designing insurance solutions beyond pure risk transfer to commercial insurers, closer in spirit to mutual insurance and rooted in the sharing economy. It also stresses the complementarity with conventional insurance, refraining to simply oppose the nowadays dominant business model to an allegedly superior new paradigm. It is quite remarkable that already many practical schemes have been developed on a trial and error basis by Insurtech entrepreneurs, backed by major (re)insurers. Few actuaries have been actively involved in this process (we discovered that a dominant market player hired a first actuary several years after it started its operations…). This book is a "must" reading for practitioners as well as students and researchers since it expands beyond classical actuarial topics. We definitely plan to adopt it in our own courses at university.

We congratulate Prof. Feng for having written this text as it occupies a niche in the range of available actuarial books that really deserved to be occupied.

Louvain-la-Neuve, Belgium — Michel Denuit
Palaiseau, France — Christian-Yann Robert
January 2023

Michel Denuit is a Professor at the UC Louvain in Belgium. Being one of the most highly cited scientists in the actuarial community, he has a wide range of research expertise in Actuarial Science, Applied Probability, Statistics, Mathematical Economics, and Operations Research. He has been recognized with numerous prizes and honors, including ARIA's Brockett-Shapiro Actuarial Journal Award 2019 and Harris Schlesinger Prize for Research Excellence 2019.

Christian-Yann Robert is a Professor at ENSAE Paris in France and Research Fellow at the Laboratory in Finance and Insurance (LFA) CREST, Center for Research in Economics and Statistics. His research interests include Extreme Value Theory and Statistics, Actuarial Theory and Practice, Statistical Finance, and Statistical Learning.

Michel and Christian have published prolifically dozens of pioneering papers on peer-to-peer insurance and risk sharing as well as a book series on machine learning and predictive analytics.

Preface

The world is never the same place with the constantly evolving Internet technology. It went from web 1.0 in the late twentieth century which was built on static information to web 2.0 in the early part of the twenty-first century with the rise of social media and user-generated content. It has been argued by many that we are entering the new stage of web 3.0 that operates through decentralized protocols. Imagine a world where products and services can be generated by anyone with the capacity around the globe and are traded in an autonomous fashion and no one or any entity has an absolute control of the process. We are already partly there. If you have ever used DeFi services such as lending on Aave and cryptocurrency exchange on Uniswap, you would recognize that new technologies have already changed how we interact with each other in the financial world. New technologies have already changed how we interact with each other in the financial world.

The insurance industry is not isolated from the evolution elsewhere in the financial world. Decentralized insurance is an encompassing term that contains many new innovations in the insurance business. A common theme of decentralized insurance schemes is that the market power that used to rest with big insurers can be returned to ordinary people. In many ways, decentralized insurance allows users to seek or provide coverage to each other without or with a reduced role of traditional insurers. Decentralized insurance has deep roots in many civilizations and the early forms of risk sharing mechanisms such as burial societies, and brotherly associations proceed today's centralized insurance industry. It is no coincidence that, with the new web 3.0 technology which is built on the premises of open, trustless, and permissionless networks, entrepreneurs return to the roots of insurance for inspiration of new business models.

As an applied scientist who has been fascinated by the fast-changing landscape of decentralized insurance, I have been interested in understanding the theoretical underpinnings of new business models. I am constantly amazed by the shared wisdom of business innovations from different cultures and markets. This is a book dedicated to fearless innovators in the insurance industry and passionate researchers in the academia. As a witness to market innovations, I want to document and summarize business logic, hoping to provide some rigorous scientific foundations and inspire practitioners and theorists for further development.

The book is intended to be used as an introduction to the technical foundation of decentralized insurance models, for advanced undergraduate students, graduate students, and practitioners. The book is self-contained and anyone with a basic knowledge of probability and statistics should be able to follow through the entire book. There is a mix of industry practices and academic theories that are presented in this book. It is by no means a comprehensive review of the literature. I take a minimalist approach to offer the essential elements and first principles so that readers can get a gist of these models without being overwhelmed by too much technicality.

Let me end with the famous quote from William Gibson: "The future is already here. It is just not evenly distributed yet." I hope to share with you the excitement of a study on decentralized insurance, which I firmly believe is in the past, the present, and the future of insurance. I hope that this book can encourage and inspire readers to innovate, experiment, and discover the next big development in the financial world.

Champaign and Hinsdale
Urbana, IL, USA
2022–2023

Runhuan Feng

Acknowledgements

This work has largely benefited from the technical support from my doctoral student, Seongyoon Kim, who has assisted in every step of the manual preparation, from the organization of raw materials, indexing, making examples and solutions, graphic illustration, to proof reading. It would have been much more difficult for me to complete this manuscript without his dedication.

I am very grateful to Prof. Michel Denuit at the Université Catholique de Louvain and Prof. Christian-Yann Robert at ENSAE Paris for their exceedingly kind and generous support in writing a foreword and pointing out typographical errors in early drafts of this book. I am also indebted to Prof. Marcin Rudź from the Lodz University of Technology for his meticulous reading of the manuscript, offering many invaluable suggestions, and correcting many typographical errors during his visit to the University of Illinois. Comments and suggestions from five anonymous reviewers are greatly appreciated as they helped improve the writing of the manuscript.

This book received generous support from several practitioners. My colleague at the University of Illinois, Tim Cardinal, extended immense support by sharing his insights from the industry perspective. Special thanks should also go to my friend, Marco Mirabella, who has always inspired me with his boundless enthusiasm and fearless endeavors in the DeFi insurance space and for his kindness to write a foreword and his encouragement for many of my projects including this book.

Last but not the least, I would also like to acknowledge the financial support from the State Farm Companies Foundation Endowed Professorship, Society of Actuaries, and Casualty Actuarial Society for several research projects that led to the creation of this book. Any opinions, findings, and conclusions expressed in this book are those of the author and do not necessarily reflect the views of the sponsors.

Contents

About the Author

Runhuan Feng is a Professor of Mathematics, Statistics, Industry and Enterprise Systems Engineering, the State Farm Companies Foundation Endowed Professor, the Director of Actuarial Science, and the Founding Director of Predictive Analytics and Risk Management at the University of Illinois at Urbana-Champaign. He is the Faculty Lead for Finance and Insurance Sector at the University of Illinois System's Discovery Partner Institute in Chicago. Runhuan is a Fellow of the Society of Actuaries and a Chartered Enterprise Risk Analyst. He co-founded the Illinois Risk Lab, which facilitates interdisciplinary activities that integrate experiential learning for students with research problems from the industry or the society. Runhuan serves as an independent consultant to many corporations and startups and provides expert testimonies to law firms for public policy assessment and actuarial analysis. His consulting work has been used by the Illinois General Assembly for pension-related legislative proposals. Runhuan's research has been recognized in the practitioners' community through his applied technical contributions and presentations as invited speakers at industry conferences. He won the Institute and Faculty of Actuaries' Geoffrey Heywood Prize in 2022. His recent interests are on business designs in decentralized finance and insurance and how distributed technologies can be best utilized to address emerging risks and societal challenges.

Chapter 1
Introduction

1.1 Primitive Risk Sharing

Risk is everywhere. Whether you engage in extreme sports like skydiving or rock-climbing or day-to-day activities as simple as driving, you face mortality risk, which is the risk of having a shortened life span and leaving behind your dependents. Risk is also throughout everyone's life from cradle to grave. Even if you live a long and healthy life, you have to deal with longevity risk, which is the risk of outliving your life-long savings and retirement income. We take risks knowingly or unconsciously in all walks of life. When you walk your dog on a rainy day, you risk being struck by lightning. When you build a sophisticated electronics system at home, you risk causing a fire that can burn down the entire house.

As risks are unavoidable, our civilizations have developed tools and strategies for managing them. For example, in agricultural societies, parents tend to have lots of children, as ways to ensure there is at least one good son or daughter who can care for them when they are old. As our societies develop social welfare systems to provide retirement incomes, the need to depend on grown-up children to manage aging-related risks is reduced. Parents worry about dying early and leaving their young children behind; they may arrange for friends to be their children's godparents, who pledge to care for the children in the absence of the parents. Or parents could buy life insurance from insurance companies which could provide a means of income for their children. As we can tell from these activities, the risks are in essence transferred and spread out to others with risk management strategies.

Risk sharing, also known as *risk distribution*, refers to the allocation of both incentives and losses for multiple parties engaged in a risk management activity with uncertain outcomes. Although the outcomes are often random and unpredictable, the financial arrangement to alleviate the negative impact can be reached before adverse outcomes may happen. Risk sharing is both an art and a science that reflects our human ingenuity throughout history.

The concept of *risk sharing* dates back centuries to many ancient civilizations. Risk sharing has always been used as an approach to ensure the fairness of wealth

R. Feng, *Decentralized Insurance*, Springer Actuarial,
https://doi.org/10.1007/978-3-031-29559-1_1

distribution in the event of disasters, and to establish accountability for economic activities involving significant risks.

1.1.1 Babylonian Laws

An early example of risk sharing appears in the oldest known legal text created during 1792–1750 BC, the Code of Hammurabi, which was a law of the ancient Babylonian civilization. The code covers a broad range of issues on many aspects of Babylonian social life. The code is casuistic and written in "if ... then ..." conditional sentences. Here are some excerpts that reflect the principle of risk sharing between a house owner and a builder for property and casualty risks.

> If a builder builds a house for a man and does not make its construction firm, and the house which he has built collapses and causes the death of the owner of the house, that builder shall be put to death.
>
> If it causes the death of the son of the owner of the house, they shall put to death a son of that builder.
>
> If it causes the death of a slave of the owner of the house, he shall give to the owner of the house a slave of equal value.
>
> If it destroys property, he shall restore whatever it destroyed, and because he did not make the house which he builds firm and it collapsed, he shall rebuild the house which collapsed at his own expense.
>
> If a builder builds a house for a man and does not make its construction meet the requirements and a wall falls in, that builder shall strengthen the wall at his own expense.

The above-stated risk sharing in the Code of Hammurabi is very simplistic and based on the principle of "an eye for an eye and a tooth for a tooth". It effectively transfers all risks of building defects and their impact on homeowners to builders. In modern societies, we have a more sophisticated home warranty and home insurance that protect the rights and responsibilities of both builders and homeowners.

1.1.2 Roman Burial Society

Ancient civilizations also developed risk sharing schemes based on social connections to cope with mortality risks. It is a time-tested financial arrangement in many societies that a group of people voluntarily pull their resources together to provide for those in need. Early examples of such risk sharing plans include the *collegium tenuiorum* in the Rome Empire, roughly translated as the association of common people, whose primary function is to assist its members to pay for funeral costs. All members are required to pay an initiation fee and monthly dues, which are deposited in a common fund and administered by the association. When a member dies, the collegium uses its common fund to assist the family of the deceased to pay for the funeral cost and possibly also to provide a bequest on behalf of the deceased. Similar functions are also common in other Roman organizations such as *collegium veteranorum*, the association of veterans, and *collegium fabrum*, the association of artisans.

Fig. 1.1 Marble inscription of the collegium of Diana and Antinous. *Source* https://followinghadrian.com/—it is published under Attribution-ShareAlike 3.0 Unported lincense (CC BY-SA 3.0), https://creativecommons.org/licenses/by-sa/3.0/legalcode

Countless such organizations played a critical role in providing a social safety net for people in all social economic classes in Roman society, especially for the poorest stratum of its population. Details of Roman collegia and their economic impact can be found in Ginsburg (1940), Perry (2006).

A well-studied example by historians is the *collegium of Diana and Antinous*. A marble inscription of the collegium was discovered in 1816 in the ruins of the city of Lanuvium in ancient Rome. The collegium was dedicated to the goddess Diana in Roman religion and Antinous, who was a Greek young man, a probable lover of the Roman Emperor Hadrian and was deified on Hadrian's orders. The Lanuvium collegium was a prominent burial society that was formed in 133 AD, three years after the death of Antinous. See in Fig. 1.1, the inscription displayed at the National Museum of Rome, Baths of Diocletian, Rome. The association consists only of men, freeborn, freedmen, and slaves. The inscription contains detailed rules and bylaws of the collegium, which in essence lay out the rights and responsibilities of its membership. Here are some excerpts on the interpretation of the inscription from Raddato (2016).

> L. Ceionius Commodus, the patron of the municipium of Lanuvium, offered the interests on 15, 000 sesterces to provide annually 800 sesterces: 400 sesterces on the birthday of Diana on August 13 and 400 sesterces on the birthday of Antinous on November 27. These financial benefactions enabled the collegium to honour Diana and Antinous and also to pay for the funerals of its members.

It was very common that these collegia receive a steady stream of revenue from the interests generated by a wealthy sponsor's large principal. Sesterce is a Roman coin and monetary unit. The inscription also spells out the membership dues, fines for missing payments, and the burial benefit offered by the association.

> Each new member had to pay an entrance fee of 100 sesterces and an amphora of good wine as well as a monthly contribution of 5 asses. If the member was up-to-date with his monthly dues when he died the association would pay his funeral expenses to the sum of 300 sesterces. However, if he failed to pay his dues for six consecutive months he would "lose the money standing in his account for the funus". Also, if a member died more than twenty Roman miles away from Lanuvium and his death was reported, the collegium would send members to take care of his funeral. If someone else took care of the funeral, the collegium was to pay this person the cost of the funeral.

Slavery was a significant but very dark part of Roman history. Under Roman laws, slaves have no personal rights but were treated as the properties of their masters, which can be traded and mistreated at the will of masters. The collegium did take into account the burial of slaves but offered inferior treatment.

> Different rules applied to slave members who were denied burial by their masters. They had the right to a fictitious funeral (funus imaginarium) which involved the cremation of a wax figure (imago) on a pyre. Meanwhile, members who committed suicides lost the right to receive funeral honours.

It is clear that the collegium offers financial arrangements to its members and mitigates the risk of funeral costs, which may be otherwise unaffordable for the family of the deceased members. The collegia were governed by local rules and the Roman Senate's regulations.

Similar social support organizations were found throughout the history of Western countries including freemasonry, medieval craft guilds, benevolent societies, etc. Many are based on the same fundamental principle of "one for all and all for one", the guiding philosophy that one should act for the benefit of a community and the community shall act for the benefit of each community member.

1.1.3 Middle East Blood Money Systems

These risk sharing plans are not unique to Western societies. Similar concepts have been in practice among tribal Arabs since the pre-Islamic period. The systems that were approved by the prophet in the Islamic region were titled al-Diyat, al-Aqilah, Ma'qil, al-Qasamah, al-Tahanud, al-Diwan, and Dhaman Khatar.

Al-Diyat, or the blood money system, among the tribal community involved compensation provided by the entire tribe to another in the case of unintentional murder. It was created as a substitute for the primitive custom of "blood called for blood" and to avoid any wars among various tribes. The *al-Aqilah* system, which translates to "the slayer's paternal relative who undertakes to pay blood money", required the members to contribute to a fund known as al-Kanz annually. This fund would in turn be used to provide compensation on behalf of any member who is subject to pay a diyat. *Ma'qil* is another system that worked similar to al-Aqilah in that all members of a tribe were required to contribute to a fund which would later be used to pay as ransom in the circumstance where someone was held hostage during a war by an enemy tribe. In the case of insufficiency of funds, related or neighboring tribes could be counted on to pay the ransom amount.

Other systems that were more specific to the occurrence of particular events included al-Qasamah, al-Tahanud, al-Muwalah, and al-Diwan. Al-Qasamah involved a contribution toward a fund that was used in the case where the family of the murder victims was unable to find the murderer. Al-Muwalah was a system involving two parties where either one would provide compensation in case anything happened to the other. Al-Tahanud was used for the purpose of food sharing while participants are traveling.

1.1.4 Granary Reserve in Ancient China

Agriculture has been vital to the growth of ancient China's economy for thousands of years. Harvest-based dynasties tend to be stable and prosperous. Famine often brings societal unrest and topples dynasties. Countless farmer uprisings throughout Chinese history had their origins in problems with food and land distribution. In the early years of the Han Dynasty, the imperial court engaged in wars to fend off its orders against the nomads in the north and the west of China. Coupled with frequent droughts, the wars consumed a large amount of resources and ordinary people no longer have sufficient food and clothing. The frequent peasants uprisings threatened the rule of the Han Empire. In recognition of these issues, Emperor Xuan of the Han Dynasty instituted a grain reserve system, known as Chang Ping Cang, across the entire country. Chang Ping Cang roughly translates to *constant leveling* or *constant unwinding* warehouse. The government bought grains at a higher price and stored them in warehouses when the market price was low. The government sold grains at a lower price and released additional supply to the market in famine years. Such a system helped average people to obtain food at a reasonable cost whether in harvest or famine years.

As the granary reserve became more abundant, the country had a better capacity to deal with natural disasters. The granary reserve system had been greatly successful to ensure social stability and was adopted by many following dynasties in China. Such a system is also a great example of risk sharing where risks of grain shortage and market disruption are spread out over the long term. This risk management strategy has been widely adopted today and used by governments around the world for national strategic reserves or stockpiles, not only for crops but also for critical resources such as medical supply, oil, and gas. The same principle is used in modern insurance that uses reserve and capital to absorb unexpected losses and to smooth out sudden shocks to revenue flows over the years.

1.1.5 Migrant Caravans in the Pakistan-India War

Risk sharing exists not only in peaceful eras but also in times of conflicts and wars. Conflicts erupted near the borders, where religious communities had coexisted, after the Indian peninsula was divided in 1947 into two independent nation-states: Hindu-majority India and Muslim-majority Pakistan. Massacres, fires, forced conversions, and kidnappings in large numbers wreaked havoc on the region. Fearing for their life, millions of people migrate across borders. Figure 1.2 shows a scene at the India-Pakistan border when people tried to catch a train to flee their homeland.

Nonetheless, there was a general consensus among the migrants that if those who hopped on at the end of trains were robbed by bandits or beggars, the rest of the caravan would all contribute to making up for their losses. Despite the horrific circumstances at times of conflict, it is a wonder of human nature that we care for one another despite having very little.

Fig. 1.2 Migrant caravan fleeing conflict zones

There are countless examples of risk sharing, which have existed as old as human civilizations. There are always needs for risk sharing in times of peace and war. As human civilizations progress, so does our ability to deal with risks.

1.2 Centralized Insurance

Most of the primitive forms of risk sharing were formed by tribal, ethnic, religious, trade groups, or other common traits. A well-known theory in anthropology suggests that there is a cognitive limit on the number of people that a person can maintain stable relationships with, called Dunbar's number. Most proponents of the theory suggest that Dunbar's number for an average person is around 150. Because these risk sharing forms are largely based on members' social networks, they are typically smaller and restricted to a limited population.

The rise of modern property insurance is often attributed to the establishment of the first insurance companies after the Great Fire of London in 1666, although marine insurance policies and markets have been developed for several centuries back. The fire made people more conscious of fire hazards and measures for recovery and repairs. Many insurance companies were set up in response to fire insurance coverage. By 1790, one out of ten houses in London was insured. Many insurance companies owned their own fire brigades. Policyholders were offered stone plates, called fire marks, to put on the front of their houses or buildings. Figure 1.3 shows a fire mark by the Hand in Hand Fire and Life Insurance Society, one of the first insurance companies started after the London fire. Before the street numbers were introduced in the 1760s, these fire marks were useful to help fire brigades identify which houses were insured.

Fig. 1.3 Fire mark

1.2.1 Centralized Models

The development of *insurance* is in essence a standardization and commercialization of primitive risk sharing. Instead of relying on the self-discipline of voluntary organizations, the agreement of mutual support is honored through contractual obligations, which are protected by civil laws. Even though each insurance contract is an agreement only between an insurer and an insured, the business model is built on the pooling of risks from hundreds of thousands of insureds under the same contract. There is in essence a common insurance fund where all policyholders' premiums are collected and out of which insurance benefits are paid to claimants. In other words, premiums collected from insureds without any loss are used by the insurer to subsidize the cost of benefits paid to other insureds with losses. It should be pointed out that insurance is at its core a *centralized model*. As shown in Fig. 1.4, the insurer, denoted by participant 0, acts as the market-maker and all payments (premium, benefit, etc.) from and to other participants $1, 2, 3, 4, 5$ transact through the insurer. It should be pointed out, however, that the transfer of payments from policyholders to claimants is behind the scene, and the process is opaque to policyholders. It is entirely in the hands of an insurer how premiums are invested and used to pay claims. The insurer is also the ultimate bearer of tail risks. If claims are higher than premiums, the insurer has to absorb the excessive costs. The basic concepts of traditional insurance are discussed in Chap. 4.

1.2.2 Pros and Cons

There are several advantages of such an organization: (1) Standardization allows insurers to use common policies, processes, and best practices across the industry. The standardization reduces administrative costs and enables insurers to sell identical

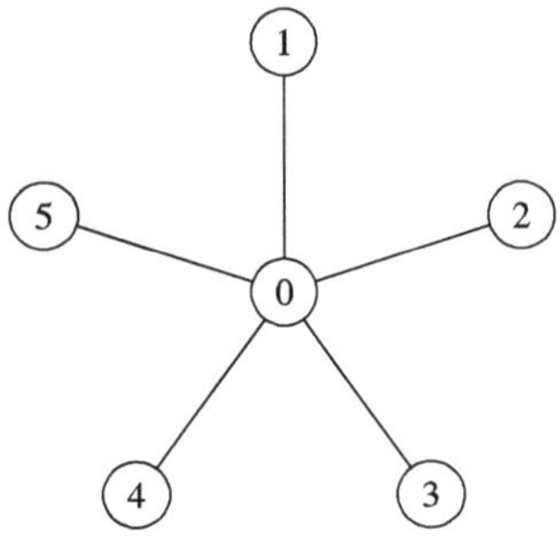

Fig. 1.4 Centralized business model

contracts to tens of thousands of policies. (2) Each policyholder deals only with an insurer, which is usually tightly regulated to meet stringent standards on their financial resources to pay for large losses in extreme events. There is little to no credit risk involved for a policyholder. (3) Insurers can take advantage of economy of scale. Selling the same contract to a large number of policyholders makes the cost of insurance predictable and manageable. While an insurer loses money on certain contracts, it is guaranteed to profit from the majority of other policies. This explains why an insurer can offer high benefit at a low premium.

However, there are also some drawbacks to the centralized model: (1) Insurance premiums are often significantly higher than the actual cost of benefits. The ratemaking often builds in profit margin, overhead, commission, the cost of solvency capital, and compliance with government regulations and other expenses. The gross premium can be multiple times the actual cost of benefit. (2) The concentration of market power rests with insurers. As they write policy terms, insurers are known to maneuver exemptions and coverage to achieve high-profit margins. High-risk individuals are often denied coverage and lack access to risk sharing mechanisms. See, for example, (Achman & Chollet, 2001). (3) There is a lack of trust in the insurance industry. Insurance operation is often opaque, and the nature of insurance is difficult to understand by average people. Many have the perception that an insurer would do anything to deny a claim. The traditional approach of commission-based sale leads to the outcome that brokers place their earnings ahead of customers' needs. A detailed discussion of the trust issue can be found in Guiso (2012).

1.3 Insurance Ecosystem

After centuries of development, the insurance industry has organized itself into a sophisticated ecosystem with interconnected sets of services. Each of these services is critically important to ensure the functioning of the insurance industry. Figure 1.5 shows a classification of insurance professionals and their roles and responsibilities in this ecosystem.

Here, we identify a number of key insurance services for which market innovations have taken place leading to discussions in later chapters.

1.3.1 Distribution

There is an old saying that "insurance is sold but not bought". Sales and marketing are critical components of insurance distribution as insurance is a service often under-appreciated by consumers, rather than a tangible commodity. Traditional insurance companies use insurance agents to sell contracts on their behalf. Insurance agents can be captive, which means that they only work for a single company, or independent, as they work for multiple companies. Many consumers also use insurance brokers who provide advisory services and assist consumers in purchasing the right policy and coverage to meet their needs. Both agents and brokers work on a commission basis at the expense of insurers. There are also general managing agents, who take on certain roles of traditional insurers, such as underwriting, pricing, and settling claims. They almost do everything else except for providing the capital to cover risks, which is a service provided by an insurer.

1.3.2 Underwriting

Underwriting is a process in which a contract is established between a policyholder and an insurer, and the latter takes on financial responsibilities in exchange for a premium from the former. The underwriting procedures vary greatly by different areas of practice. For example, health insurance may require a policyholder to submit their medical records. Auto insurance may only require a policyholder to fill out a questionnaire regarding the vehicle's make, year, model, etc. Professional underwriters evaluate the risk involved with each applicant and the exposures for the insurer. The responsibilities of an underwriter include determining whether to accept or decline an application, how much surcharge to impose if the risk is deemed substandard, and what exemptions should be stipulated to avoid excessive exposure. Despite the efforts of insurance companies to standardize the underwriting procedure by developing underwriting manuals, it still requires years of experience for an underwriter to be able to make the right judgment calls. Automated underwriting has been introduced for many applications, although the technology is in its infancy for human-related coverage that calls for nuanced judgment.

1.3.3 Pricing

Pricing, also known as ratemaking, refers to the process of setting the premium at which an insurance contract is sold. Insurance prices typically depend on the probability of loss, the operating expenses, the competitiveness of the products in

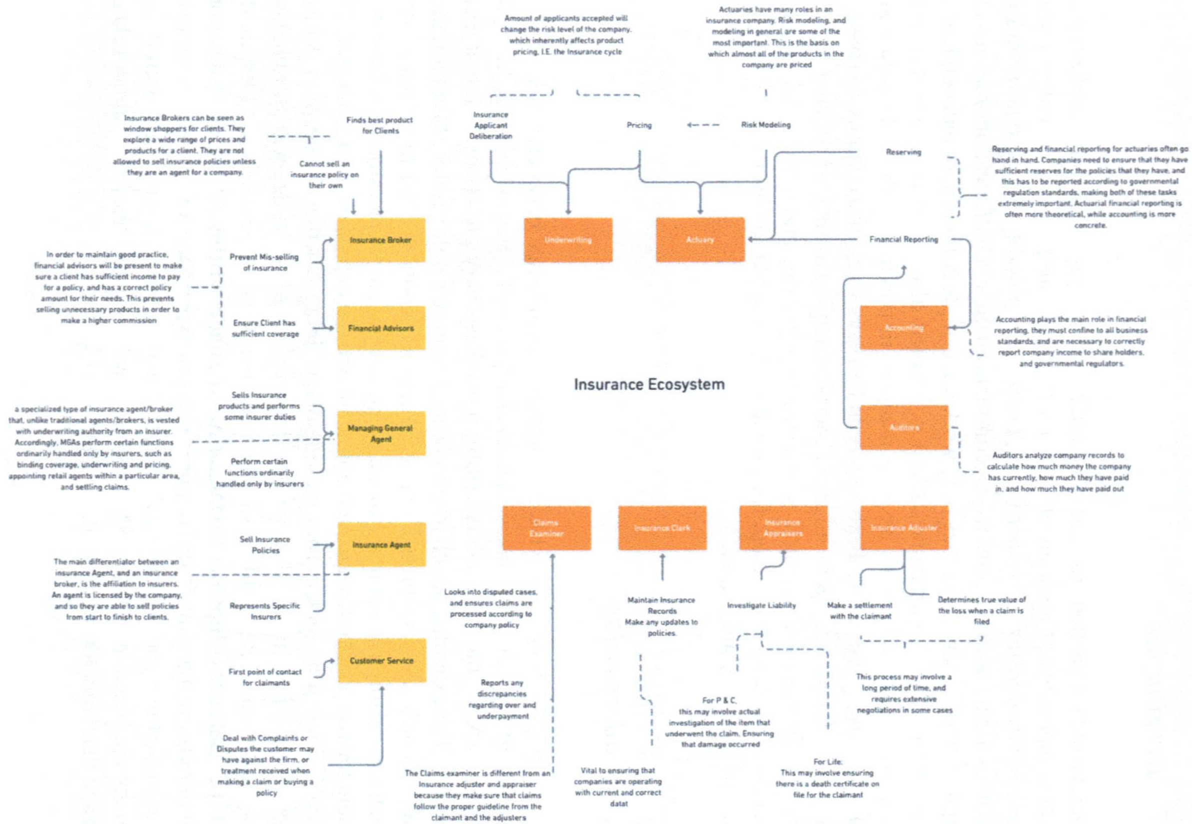

Fig. 1.5 Insurance ecosystem and professionals

the market, and the profit margin of an insurer. The traditional insurance industry uses professionally trained actuaries to perform actuarial analysis and capital management. Actuaries usually price insurance products based on mathematical and statistical models on historic claims data for cost prediction and profit/loss analysis.

1.3.4 Asset and Liability Management

Insurers, especially life, annuity, and pension, tend to have short-term assets and long-term liabilities. There are often substantial financial risks involved due to the mismatch between assets and liabilities. Actuaries develop many strategies to invest in assets with varying maturities to match the schedule of liabilities due. They perform simulations and stress testing on how their assets and liabilities behave under adverse economic scenarios.

1.3.5 Claims

Claims processing refers to the steps that an insurer takes to handle a claim, including collecting information about claims, assessing the eligibility and the extent of damage or losses, investigating disputed claims and settling payments, etc. For example, property insurers use professional adjusters to review claims, inspect the damage, and assess the appropriate amount of compensation. If a claim is in dispute, an insurer and the policyholder may submit the claim to a third-party claims appraiser. Both parties make good faith attempts to resolve the dispute before resorting to litigation. If an insurer defects a potential fraud such as falsified claims or staged accidents, the insurer may hire a claims investigator to perform an investigation like a private detective.

In recent decades, there have been new developments in parametric insurance, or index-based insurance, for which payouts are based on parameterized triggers. For example, a payment from earthquake insurance may be automatically approved if an earthquake of a certain magnitude occurs, regardless of the actual damage to a policyholder. The main benefit of parametric insurance is that the trigger is easy to verify and the insurer can provide immediate disaster relief. It avoids a potentially lengthy process of claims processing.

1.3.6 Governance

Traditional insurers are usually either stock companies or mutual companies, depending on their ownership structure. The modern insurance industry started with mutual insurance in the 1600s. A mutual insurance company is owned by its policyholders to serve the needs and protect the benefits of policyholders and their beneficiaries. A stock insurance company is a corporation owned by stockholders with the objective to maximize its equity value. Policyholders do not participate in the profits or losses of stock companies.

1.3.7 Regulation

As insurance business is in general centralized and each insurer serves hundreds of thousands of policies, an insurer's ability to honor its financial obligations is of concern to governments in order to protect the interests of the public. Insurance regulators issue insurance licenses, require insurers to submit annual financial reports, examine their market conducts, exert price control to avoid unfair and unethical practices of pricing, ensure sufficient solvency capital requirements, etc.

1.4 Decentralized Insurance

Decentralized insurance is a broad umbrella that covers a wide range of insurance business models that reduce or eliminate the role of market intermediaries in the insurance ecosystem. It is sometimes also referred to as *disintermediated insurance*, as such practices bypass or cut out the insurer as a middleman. While the two terms decentralization and disintermediation have different connotations, the reduction of the insurer's role as the central authority and the intermediary represents both concepts, and the two terms have been used interchangeably in the literature.

1.4.1 Rise of Sharing Economy

The recent decentralization movement started with the rise of sharing economy. With the adoption of the Internet and mobile phones, users can directly interact with each other and receive or provide goods or services without traditional distribution channels. One example is Uber, a platform that connects passengers with drivers. Unlike traditional taxi companies, Uber owns no fleet of automobiles. It simply offers a match-making service that enables people with idle cars to provide services to those in need of a ride at a low cost. Similarly, the hotel business is disintermediated by Airbnb, which offers users a lower price by allowing average people to offer their homes to travelers. In addition to transportation and hospitality, many other industries such as media and education are also increasingly moving toward disintermediated business models. For example, instead of relying on publications by leading authorities, most people nowadays use Wikipedia or other digital encyclopedias to look for information. Similarly, the new generation of consumers tends to use Twitter, Facebook, Instagram, and TikTok for information rather than traditional news outlets. The body of knowledge and information on these websites are collectively created, updated, and maintained by a global community of users. While there is certainly

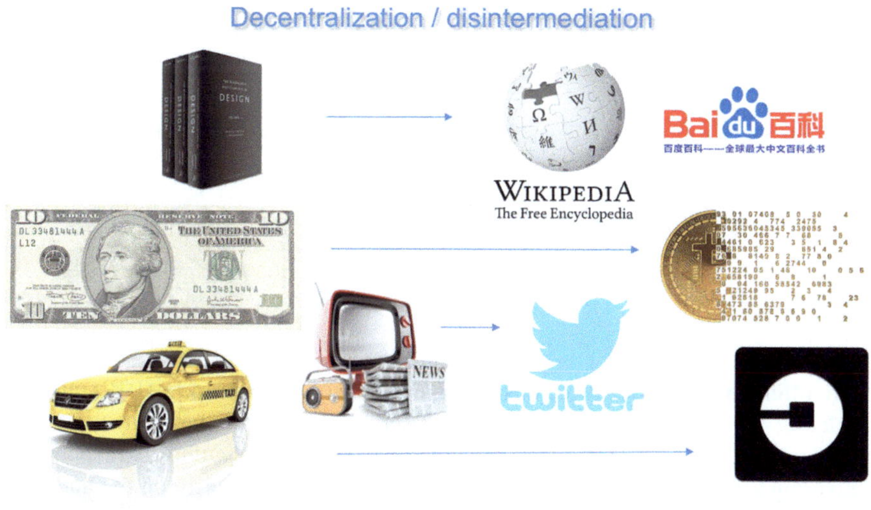

Fig. 1.6 Decentralization in sharing economy

misinformation or disinformation on the Internet, there is no central authority that can dictate how facts and news are recorded and reported (Fig. 1.6).

Against the backdrop of decentralization in many parts of our societies, it is not surprising that decentralized business models start to emerge in the financial industry as well. One example is the rise of cryptocurrencies like Bitcoin, which is a digital, decentralized peer-to-peer payment system with no involvement of any monetary authority. Another example is the booming peer-to-peer lending business, as an alternative financing method. As the name implies, peer-to-peer lending enables individuals to obtain loans directly from other individuals, cutting out the financial institution as the middleman. The removal of financial intermediaries such as banks is argued to reduce cost and improve efficiency.

1.4.2 Decentralized Models

The insurance industry is no exception. It is very natural that many new business models emerge as a revival of early forms of risk sharing with modern enabling technologies. As we shall study in the rest of this book, many business models can find their roots in primitive risk sharing. Instead of transferring risks to a centralized intermediary like an insurer, many models adopt strategies to spread out the risks directly among a network of users. In Chap. 5, we shall introduce and examine a range of decentralized insurance business models, including online mutual aid in East Asia, takaful in the Middle East, peer-to-peer insurance in the West, and catastrophe risk pooling in the Caribbean and Central America countries. Other decentralized insurance practices include health share plans, such as Christian Health Ministries, longevity risk pooling, and tontines.

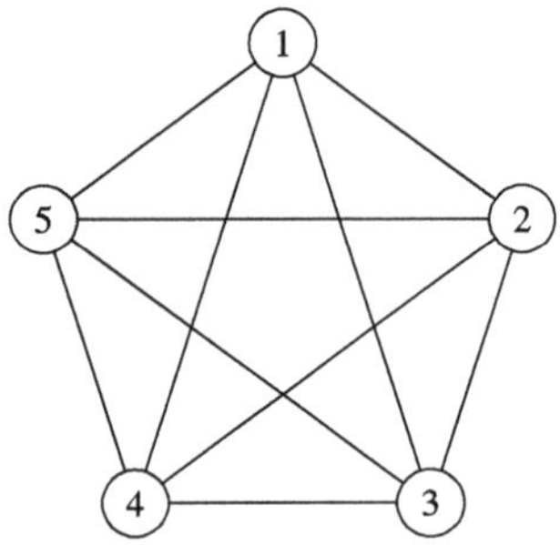

Fig. 1.7 Peer-to-Peer business model

To offer a brief preview, we consider an example of a peer-to-peer (P2P) business model as opposed to the traditional client-server model. As shown in the client-server model in Fig. 1.4, clients make requests and the central authority responds by delivering services. The functioning of the system relies primarily on the authority and the capacity of the server. In contrast, Fig. 1.7 visualizes a peer-to-peer model which requires no central authority and instead is made up of all peers with rights and obligations. Users on the P2P platform may utilize social media to invite family and friends to form their mutual support groups. As depicted in Fig. 1.7, friends or family relatives (peers $1, 2, 3, 4, 5$) with the same insurance needs enter into a mutual agreement that if anyone suffers a loss then the loss shall be split and carried by all members in the group. The clear advantage of such a model is that it avoids the use of an insurer as a market intermediary. Even though there are market facilitators that provide the platforms for onboarding and serving clients, the facilitators do not take on any risk and use pass-through mechanisms. We shall discuss in great detail the concepts of aggregate risk pooling and P2P risk exchange in Chaps. 6 and 7, respectively.

In recent years, there is also an emergence of new business models built on blockchain technology, which is also known as distributed ledger technology. The new technology creates unprecedented opportunities for entrepreneurs to offer innovative services. A subset of these services, called *decentralized finance*, or DeFi for short, enable users to receive and provide financial services to each other without traditional financial institutions. Most of DeFi services with large market capitalization are lending/borrowing and crypto exchanges, the counterparts of banking and security exchanges in traditional finance. Although still in its infancy, DeFi insurance is a growing class of insurance services that utilize blockchain technology to decentralize many components of the insurance ecosystem, such as pricing, underwriting, and capital management. We shall provide an overview of blockchain technology, decentralized finance, and DeFi insurance in Chap. 9.

To wrap up this introduction, we show where this book fits in the field of insurance and risk management. Figure 1.8 shows the connections between traditional insurance, decentralized insurance, and blockchain technology.

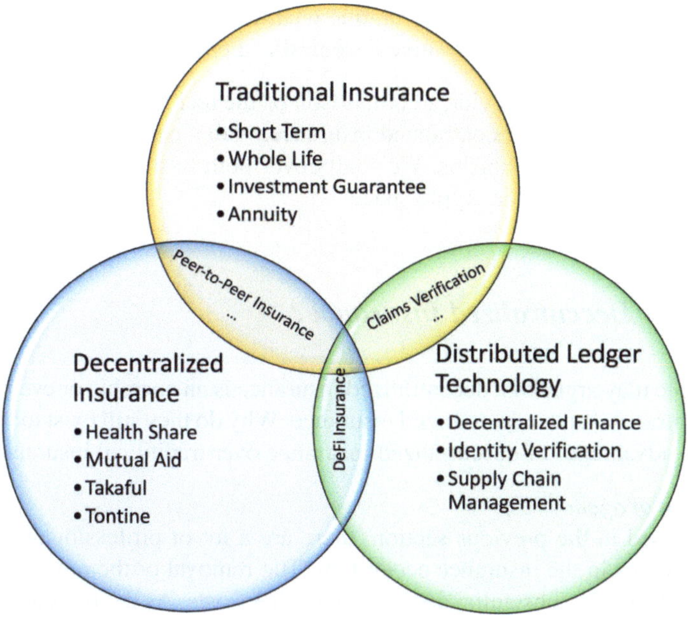

Fig. 1.8 Decentralized insurance versus traditional insurance

- *Traditional versus Decentralized Insurance*
  As explained earlier, decentralized insurance has existed in various forms for a long time in parallel to the modern insurance industry. However, there are overlaps between the two categories of insurance. For example, peer-to-peer insurance has been used as supplement to traditional insurance. Most of the peer-to-peer insurance products that exist in Europe and North America are automobile, healthcare, and renters' insurance. Policyholders typically have traditional insurance products with very high deductible. The peer-to-peer pools allow them to create a common fund where small claims below the deductibles can be covered
- *Traditional Insurance versus Distributed Ledger Technology*
  The blockchain technology has much broader applications beyond finance and insurance, such as identity verification and supply chain management. Readers should be reminded that there are many blockchain applications that enable traditional insurers to collaborate with each other. For example, RiskStream Collaborative is a blockchain consortium, managed by the American Institute for Chartered Property Casualty Underwriters, that develops private chain solutions to enable insurers to exchange personally identifiable information data in compliance with regulations to prevent fraud.
- *Decentralized Insurance versus Distributed Ledger Technology*
  Cryptocurrency fraud and scams that steal investors' crypto funds are growing pains with the fast development of crypto assets and transactions. Smart contract insurance, which can aid investors in recovering losses, has been developed in

response to market demand. The unique characteristics of the DLT also enable entrepreneurs to develop innovative decentralized business models for insurance.

This book is intended to provide a comparison of the technical foundations of both traditional insurance and decentralized insurance, with a particular focus on decentralized risk sharing mechanisms. We shall cover both business models in the real world as well as those in the crypto space.

1.4.3 Why Decentralized Insurance?

Many people may argue that decentralized insurance is an anarchic or even primitive form of centralized and industrialized insurance. Why do they still exist today? There are several advantages of decentralized insurance over traditional insurance.

- *Reduction of agency cost*
  As discussed in the previous section, there are a lot of professionals serving as intermediaries in the insurance ecosystem. The removal or the reduction of their roles can lead to a substantial decrease in agency costs. As the risks are no longer concentrated on a single authority, there is less need for heavy capital requirement to absorb "black swan" events, leading to less compliance cost with insurance regulations.
- *Pendulum of uncertainty*
  Insurance is a business dealing with uncertainty. When the risks are self-retained, individuals have to face the uncertainty of potential severe losses. When the risks are transferred from policyholders to an insurer, there is little to no risk left for each policyholder as they often pay fixed premiums in exchange for fixed benefits. If we can place the degree of uncertainty on a spectrum (see Fig. 1.9), self-insurance is on one end of the extreme and full insurance is on the other end. The same concept goes for annuities and pensions. When individuals save for retirement, they face the uncertainty of outliving their savings. They could gain peace of mind by purchasing annuities for steady income. But all risks are instead taken on by an insurer. Like annuities, defined benefit retirement plans offer guaranteed incomes to retired employees and the burden of investment risks is borne by the employer. In analogy to personal savings, defined contribution plans only require employers to make contributions into employees' accounts and leave all investment risks to employees. Over the past few decades, there has been a major shift of employee benefits around the world moving from defined benefit plans to defined contribution plans in order to reduce their financial uncertainty. When the pendulum swings too far, it has to come back in the other direction. There have been debates and discussions in the retirement industry for new hybrid plans such as target benefit and collective defined benefit that require plan sponsors to take on some risks (like traditional insurance and annuity) but distribute some uncertainty among participants (similar to decentralized insurance and annuity). Decentralized insurance

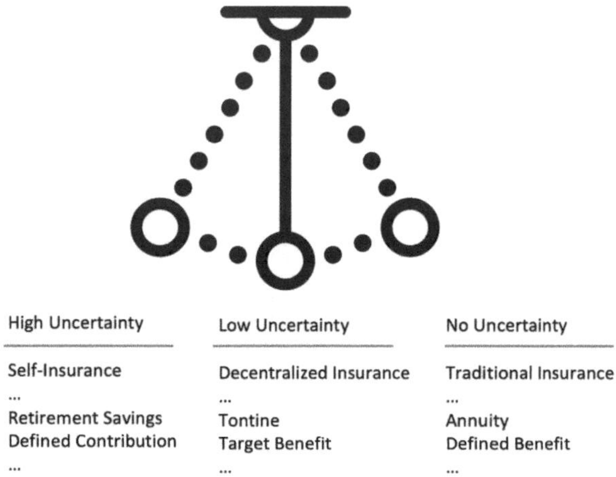

Fig. 1.9 Pendulum of uncertainty

is in essence a mechanism to allow participants to share risks with each other, in which some risks may be partially passed on to a central entity. It reduces risks for every participant due to diversification. While it is not as risky as self-insurance for each participant, it is also less risky for the central entity which would otherwise take on all risks in a traditional full insurance approach. Therefore, decentralized insurance can be viewed as a middle point on the spectrum of uncertainty.

1.4.4 Structure of the Book

The focus of this book is to introduce, analyze, and compare various new decentralized insurance business models in practice as well as theoretical models in the academic literature all developed in recent decades. We hope to provide a systematic approach to exploit the theory underpinning decentralized insurance and to appreciate the wisdom of practitioners and academics in their different perspectives.

As we follow the logic to first present key components of traditional insurance and then move on to their counterparts in decentralized insurance, many chapters and sections are interconnected as shown in Fig. 1.10. This chapter offers the background on decentralized business models, which are connected to decentralized insurance models in later chapters, and discusses the insurance ecosystem, of which some components are decentralized and discussed in Chap. 9. Chapter 2 lays out the technical foundation for risk modeling and assessment at an introductory level. The modeling tools are used to describe both traditional insurance in Chap. 4 and decentralized insurance in Chaps. 6 and 7. An overview of the fundamentals of risk management and insurance is provided in Chapter 3. We shall illustrate in Chaps. 4, 8, and 9 that these risk management tools have been mixed and utilized for various types of insurance business models. Many economic concepts such as utility maximization and Pareto optimality are used in Chaps. 4, 6, and 7. We discuss basic concepts of pric-

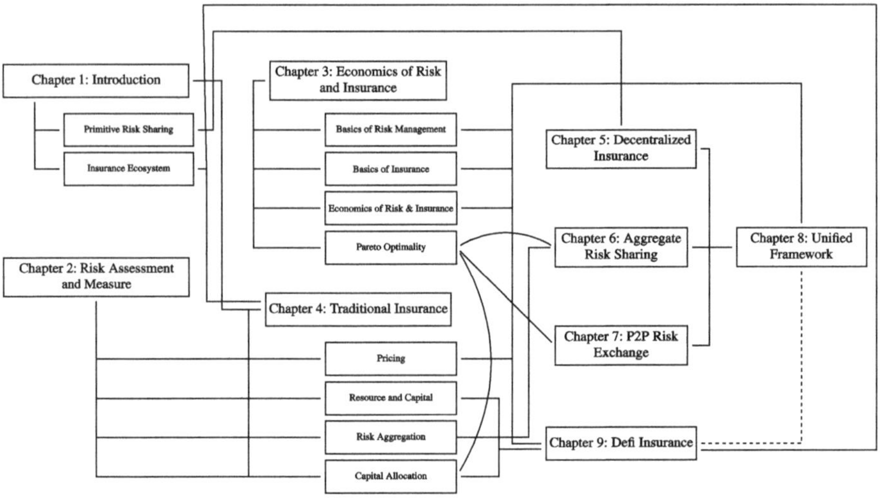

Fig. 1.10 Logical connections in the book

ing, reserving, risk aggregation, capital management, and allocation in the context of traditional insurance in Chap. 4. The same concepts are extended to decentralized settings in later chapters. For example, aggregate risk sharing methods in Chap. 6 are very similar to that for traditional insurance; fair risk sharing rules in Chaps. 6–8 are based on the same idea of differential pricing in traditional insurance. Some risk aggregation and capital allocation methods are shown in the context of traditional insurance in Chap. 4 have been adopted by DeFi insurance businesses and are further discussed in Chap. 9. While Chap. 5 collects a range of business models developed in the physical world, Chap. 9 is dedicated to those that emerged with blockchain technology. Chapter 8 provides a unified framework for understanding decentralized insurance schemes, of which Chaps. 5 and 7 are special cases and Chap. 9 can be viewed as an extension.

1.4.5 Closing Remarks

We hope this introduction convinces the readers that decentralized insurance is both historic and rapidly involving with modern society. The goal of the book is to share with readers the excitement of deciphering the "secret sauce" of decentralized insurance. Admittedly, the book is far from comprehensive to cover all the business innovations and academic theories. For example, most of the discussions in this book are restricted to one-period models. We shall address multi-period models and decentralized annuities in a sequel. What is contained in this book is merely a tip of the iceberg on this topic. As an old saying by Lao Tzu goes, "a journey of a thousand miles begins with a single step." Let the journey begin.

Chapter 2
Risk Assessment and Measures

This chapter offers an overview of fundamental tools for quantifying, measuring, and assessing risks. Univariate distributions are commonly used to describe the randomness of particular risks to be quantified and modeled. Risk measures are summary statistics that portray various aspects of risks. They are often used by financial institutions as quantitative bases to set risk management policies. Risk measures can be further used to establish the ordering of risks for the purpose of comparison. The ordering of risks is used in later chapters as a gauge for the effectiveness of risk reduction strategies. As risks are often interconnected, multivariate distributions are critical tools for measuring and understanding their relationships. Dependence measures are also introduced as summary statistics that characterize the strength of dependence between risks.

We assume that readers have some basic knowledge of random variables, which are usually discussed in the first course on probability and statistics. This chapter merely serves as a recap of key elements as building blocks for risk modeling and analysis.

2.1 Univariate Modeling

Risk is often interpreted as the uncertainty of potential losses. In the context of insurance, we often consider undesirable financial outcomes, such as the loss of property values for property owners resulting from catastrophic events, and loss of income for a worker due to a disability. Hence, we use the terms risk and loss interchangeably in this book.

The most common approach to quantify and assess risks is to model them by random variables. In simple words, a random variable is a quantity that takes on a range of values according to some known probabilities. These probabilities represent our understanding of the nature of randomness for a particular risk. It tells us with

R. Feng, *Decentralized Insurance*, Springer Actuarial,
https://doi.org/10.1007/978-3-031-29559-1_2

what odds a loss can happen at which size. For example, a statement such as “category 5 hurricane systems with potential damage of more than \$5 million dollars develop only about once every three years on average in the Atlantic region” is a description of properties of a loss distribution.

2.1.1 Loss Distribution

A random variable is always associated with a set of probabilities. In mathematical terms, these probabilities are defined by a so-called probability measure, denoted by $\mathbb{P}$ as in most books. Roughly speaking, it measures the probability of any event involving the random variable under consideration. A random variable is fully characterized by a distribution function.

Definition 2.1 The (cumulative) distribution function, denoted by F_X, of a random variable, X, is given by

$$F_X(x) = \mathbb{P}(X \leq x).$$

The survival function, denoted by $\overline{F}_X$, is given by $\overline{F}_X(x) = 1 - F_X(x)$.

A distribution function always has the following properties: (1) It is non-decreasing, i.e. $F_X(x) \leq F_X(y)$ if $x \leq y$. Intuitively speaking, this is because the probability $F_X(y) - F_X(x) = \mathbb{P}(x \leq X \leq y)$ can never be negative. (2) It is also right-continuous, i.e. $\lim_{h\to 0, h\geq 0} F_X(x+h) = F_X(x)$. This property is by convention to avoid ambiguity at the point of jump. (3) As x approaches ∞ or $-\infty$, the limits of the distribution function should be 1 or 0, respectively, i.e. $\lim_{x\to\infty} F_X(x) = 1$, $\lim_{x\to-\infty} F_X(x) = 0$. The probability of having all possibilities of loss must be one.

The distribution function F is said to be absolutely continuous if there exists a function f such that

$$F_X(x) = \int_{-\infty}^{x} f_X(y)\mathrm{d}y.$$

The function f_X is called the density function. When such a function exists, we say X is a continuous random variable.

Example 2.1 Consider a standard normal random variable (with mean 0 and variance 1), denoted by N. Its distribution function is defined to be

$$\Phi(x) = \mathbb{P}(N \leq x) = \int_{-\infty}^{x} \frac{1}{\sqrt{2\pi}} e^{-t^2/2} dt.$$

A general normal random variable can be written as a linear transformation of the standard normal random variable. For example, a normal random variable with mean μ and variance σ^2 can be written as $X = \mu + \sigma N$. Then its distribution function can be obtained by

$$F(x) = \mathbb{P}(X \le x) = \mathbb{P}(\mu + \sigma N \le x) = \mathbb{P}\left(N \le \frac{x-\mu}{\sigma}\right) = \Phi\left(\frac{x-\mu}{\sigma}\right).$$

2.1.2 *Quantile Functions*

An equivalent way to describe a random variable is to use the inverse of its distribution function, also known as the quantile function. As we shall explain below, there could be some ambiguity arising from the definition of an inverse. Therefore, we have to use some technicality to provide a precise definition.

Definition 2.2 Consider a loss random variable X whose distribution function is denoted by F. Given a probability level $p \in (0, 1)$, the quantile function is defined as

$$F_X^{-1}(p) = \inf\{x : F(x) \ge p\} = \sup\{x : F(x) < p\},$$

where $\inf A$ is the infimum of a set A and $\sup A$ is the supremum of the set.

In simple words, $F_X^{-1}(p)$ is the smallest threshold exceeded by X with the probability of $1 - p$. There are essentially three different cases that one needs to consider when computing the quantile function. In general, the graph of an inverse function is given by the mirror image of the graph of the original function by the line $y = x$, as shown in Fig. 2.1. The three cases are represented by the asterisks in the figure. When the value p corresponds to a flat piece on the graph of the cumulative distribution function, all points on the x-axis correspond to the same function value. Therefore, ambiguity arises when one attempts to define the inverse of such a function value. In such a case, $F_X^{-1}(p)$ is always defined to be the left endpoint of the interval on which the flat piece appears. It is also clear from Fig. 2.1 that $F_X^{-1}(p)$ as a function of p is left-continuous.

There is an equivalence of inequalities involving inverse functions that is very useful in applications.

$$F_X^{-1}(p) \le x \Longleftrightarrow p \le F_X(x). \tag{2.1}$$

The proof of this result is relegated to Appendix 2.A of this chapter.

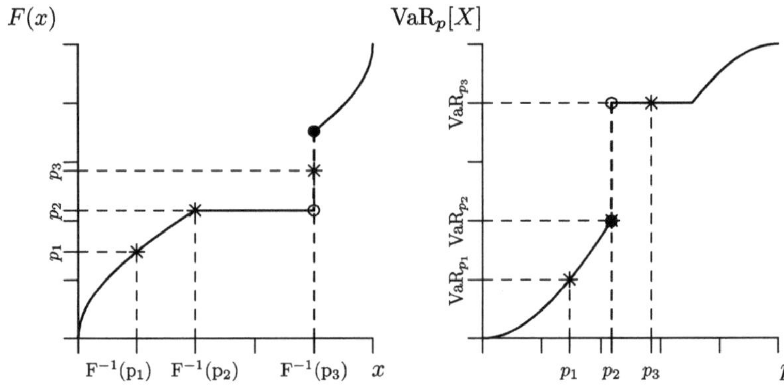

Fig. 2.1 Value-at-risk and cumulative distribution function

Example 2.2 Consider any real-valued random variable X with distribution function F_X and a uniform random variable U defined on [0, 1]. Then X has the same distribution as $F_X^{-1}(U)$. This is because $\mathbb{P}(F_X^{-1}(U) \leq x) = \mathbb{P}(U \leq F_X(x)) = F_X(x)$ as a result of the equivalence in (2.1). This result offers an approach to generate a random sample of X by applying F_X^{-1} to the uniform random sample. This is called the inverse function method for a random variable generation.

Example 2.3 If a random variable X has a continuous distribution function F_X, then $F_X(X)$ has the same distribution as U where U is a uniform random variable on [0, 1]. Observe that $\mathbb{P}(F_X(X) \geq u) = \mathbb{P}(F_X^{-1}(u) \leq X) = 1 - F_X(F_X^{-1}(u))$, due to the equivalence in (2.1). If F_X is a continuous function, then $\mathbb{P}(F_X(X) \geq u) = 1 - u$, which indicates that $F_X(X)$ is also uniformly distributed.

2.2 Risk Measures

Since risks are often quantified by random variables, it is important to develop metrics to assess the impact of taking on risks for a business. The idea of using a metric to summarize information is very prevalent in everyday life. For example, the highest and lowest temperatures in daily weather forecasts are metrics of temperature random variables; a university ranking offers a metric to measure the quality of university education in comparison to its peer institutes. These simple metrics allow an average person to prepare for the weather without understanding complicated atmospheric chemistry and physics, and give a prospective student a sense of the relative competitiveness of a university saving them from time-consuming research on their own. In the same spirit, a risk measure is intended to offer a summary of the nature of a risk random variable in order for a decision maker to take an action with regard to the risk.

Definition 2.3 A risk measure is a mapping from a random variable X to a real number $\rho(X)$.

Different university rankings may give different ranks to a university. Meteorologists from various TV stations can have different forecasts for the same area. Likewise, different risk measures reflect different aspects of a risk. For example, the mean measures the central tendency of randomness, whereas the variance measures how far the outcomes of a risk are spread out from its mean. Assume that X is a continuous random variable. Then, its mean and variance are defined by

$$\mathbb{E}[X] = \int_{-\infty}^{\infty} x f_X(x)\,\mathrm{d}x;$$
$$\mathbb{V}[X] = \int_{-\infty}^{\infty} (x - \mathbb{E}[X])^2 f_X(x)\,\mathrm{d}x,$$

where f_X is the probability density function of X.

While mean and variance are useful risk measures, there are other risk measures that are more informative for the purpose of risk management. A common strategy for risk management is to quantify and assess the likelihood and severity of losses, which can be used later to develop mitigation methods. Common risk measures that can characterize likelihood and severity include Value-at-Risk (VaR), Tail-Value-at-Risk (TVaR), etc.

2.2.1 *Value-at-Risk*

In many areas of risk management practice and regulation, the concept of *Value-at-Risk* (VaR) is introduced to measure the likelihood of severe losses to a business. It is the same as the quantile function, i.e. $\mathrm{VaR}_p[X] = F_X^{-1}(p)$.

Example 2.4 Consider the value-at-risk of a normal random variable X with mean μ and variance σ^2. As shown in Example 2.1, the distribution function in this case is continuous and strictly increasing, then the value-at-risk is the ordinary inverse function, which can be written as

$$\mathrm{VaR}_p[X] = \mu + \sigma\Phi^{-1}(p), \tag{2.2}$$

where Φ^{-1} is the inverse function of Φ.

Example 2.5 Consider two loss random variables L_1 and L_2 from an insurer's point of view. The random variable L_1 is exponentially distributed with the cumulative distribution function given by

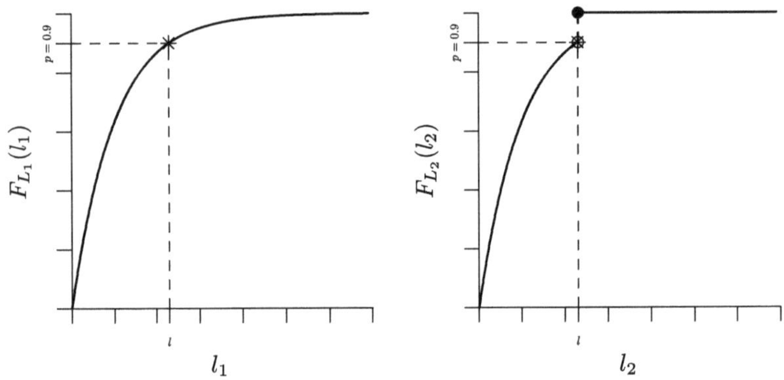

Fig. 2.2 Value-at-risk and cumulative distribution function

$$F(x) = 1 - e^{-\lambda x}, \qquad x \geq 0.$$

It is clear that the value-at-risk of L_1 is given by

$$\text{VaR}_p[L_1] = -\frac{1}{\lambda}\ln(1-p), \qquad p \in (0, 1).$$

The random variable $L_2 := \min\{L_1, u\}$, which represents the loss L_1 subject to a policy limit of $u = \text{VaR}_{0.9}(L_1) = \ln 10/\lambda$.

As indicated in Fig. 2.2, one can show by definition that $\text{VaR}_p[L_1] = \text{VaR}_p[L_2]$ for all $p \leq 0.9$ and $\text{VaR}_p[L_1] > \text{VaR}_p[L_2]$ for all $p > 0.9$. Even though the unbounded loss L_1 can be more dangerous than the capped loss L_2, the insurer would only set aside the same amount of capital if $\text{VaR}_{0.9}$ is the chosen risk measure as the basis of capital requirement. This shortcoming of VaR is often criticized in financial applications and gives an argument to proponents of alternative risk measures such as tail-value-at-risk, to be introduced in the next section.

An interesting fact is that $\text{VaR}_{1-\epsilon}[X]$ is a solution to the following minimization problem:

$$\min_{c \in \mathbb{R}} \left\{ \mathbb{E}[(X-c)_+] + c\epsilon \right\}, \qquad 0 < \epsilon < 1, \tag{2.3}$$

where $(x)_+ = \max\{x, 0\}$. The proof can be found in Appendix 2.B. Here, we provide an economic interpretation of the quantity in curly brackets. Note that if c is interpreted as an insurer's total capital, then the first term is the expected value of shortfall in the capital after taking on the loss X. Of course, if the sole purpose is to minimize the expected shortfall, then the capital c should be set as high as possible. Note, however, holding capital incurs a cost to an insurer, as the allocated fund is unavailable for any other corporate purposes. The second term is to penalize the use of capital by imposing the cost of capital at the rate of ϵ per unit, say the borrowing

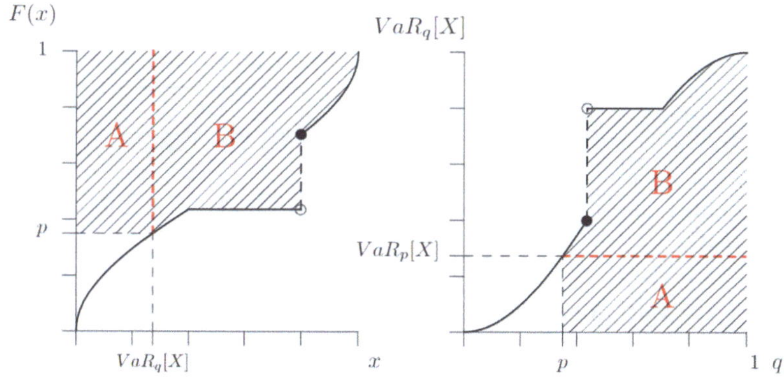

Fig. 2.3 Value-at-risk and cumulative distribution function

interest rate of 5% per unit. The greater the amount of capital, the higher the cost. The goal of the minimization problem (2.3) is to strike a balance between minimal capital shortfall and minimal cost of capital.

Note that, if we let $p = 1 - \epsilon$, then the minimal value in (2.3) is given by

$$\mathbb{E}[(X - \mathrm{VaR}_p[X])_+] + \mathrm{VaR}_p[X](1 - p). \tag{2.4}$$

2.2.2 *Tail-Value-at-Risk*

As demonstrated in Example 2.4, a shortcoming of VaR as a risk measure is that it does not say much about the severity of losses beyond the threshold. A remedy to this problem is provided with the concept of tail-value-at-risk.

Definition 2.4 The Tail-Value-at-Risk (TVaR) of a loss random variable X at a probability level $p \in (0, 1)$ is defined by

$$\mathrm{TVaR}_p[X] = \frac{1}{1-p}\int_p^1 \mathrm{VaR}_q[X]dq.$$

Recall that the graph of the inverse function $\mathrm{VaR}_p[X]$ is a reflection of the distribution function $F_X(x)$ across the line $y = x$. As shown in Fig. 2.3, the shaded areas, A and B, in the two graphs are the same, which implies that

$$\int_p^1 \mathrm{VaR}_q[X]dq = \underbrace{\mathrm{VaR}_p[X](1-p)}_{A} + \underbrace{\int_{\mathrm{VaR}_p[X]}^{\infty} \overline{F}_X[q]dq}_{B}$$
$$= \mathrm{VaR}_p[X](1-p) + \mathbb{E}[(X - \mathrm{VaR}_p[X])_+].$$

The last equality can be obtained using an argument similar to that for (2.17). Then it is clear that

$$\mathrm{TVaR}_p[X] = \mathrm{VaR}_p[X] + \frac{1}{1-p}\mathbb{E}[(X - \mathrm{VaR}_p[X])_+].$$

Recall the minimization problem (2.3) and its minimum value (2.4), which implies that

$$\mathrm{TVaR}_p[X] = \min_{c\in\mathbb{R}} \left\{ \frac{1}{1-p}\mathbb{E}[(X-c)_+] + c \right\}. \tag{2.5}$$

Example 2.6 Show that the TVaR of a normal random variable X with mean μ and variance σ^2 is given by

$$\mathrm{TVaR}[X; p] = \mu + \sigma\frac{\phi(\Phi^{-1}(p))}{1-p}, \qquad p \in (0, 1),$$

where Φ is the cdf of a standard normal random variable and ϕ is its pdf.

Proof Let $X \sim \mathrm{N}(\mu, \sigma^2)$, $Z \sim \mathrm{N}(0,1)$. Then $X = \mu + \sigma Z$, and

$$\begin{aligned}
\mathrm{TVaR}(X; p) &= \frac{1}{1-p}\int_p^1 \mathrm{VaR}(X; \lambda)\mathrm{d}\lambda \\
&= \frac{1}{1-p}\int_p^1 \mathrm{VaR}(\mu + \sigma Z; \lambda)\mathrm{d}\lambda \\
&= \frac{1}{1-p}\int_p^1 F^{-1}_{\mu+\sigma Z}(\lambda)\mathrm{d}\lambda \\
&= \frac{1}{1-p}\int_p^1 \mu + \sigma\Phi^{-1}(\lambda)\mathrm{d}\lambda \\
&= \mu + \frac{\sigma}{1-p}\int_p^1 \Phi^{-1}(\lambda)\mathrm{d}\lambda.
\end{aligned}$$

Let $z = \Phi^{-1}(\lambda)$, $\lambda = \Phi(z)$, and $\mathrm{d}\lambda = \phi(z)\,\mathrm{d}z$, then

$$\begin{aligned}\text{TVaR}(X;p) &= \mu + \frac{\sigma}{1-p}\int_{\Phi^{-1}(p)}^{\infty} z\phi(z)\mathrm{d}z \\ &= \mu + \frac{\sigma}{1-p}\int_{\Phi^{-1}(p)}^{\infty} (-\phi(z))\,\mathrm{d}z \\ &= \mu + \sigma\frac{\phi(\Phi^{-1}(p))}{1-p}.\end{aligned}$$

2.2.3 Coherent Risk Measure

While Definition 2.3 offers little restriction on what constitutes a risk measure, there are certain common properties of risk measures that guide a decision maker's search for a desirable risk measure. A concept well-accepted in the financial literature is *coherent risk measure*, which was first introduced by Artzner et al. (1999).

Definition 2.5 A risk measure ρ is said to be coherent if it satisfies the following properties for all random variables X, Y:

- (Translativity) $\rho[X + c] = \rho[X] + c$ for any constant c;
- (Positive homogeneity) $\rho[cX] = c\rho[X]$ for any constant $c > 0$;
- (Subadditivity) $\rho[X + Y] \leq \rho[X] + \rho[Y]$;
- (Monotonicity) If $\mathbb{P}(X \leq Y) = 1$, then $\rho[X] \leq \rho[Y]$.

It is helpful to keep in mind that risk measures are often used for regulatory purposes. The primary objective of a regulator is to ensure market stability by requiring financial institutions in its jurisdiction to hold sufficient funds, e.g. reserve and capital, to cover its liability even under adverse economic conditions. The quantification and assessment of risk through a risk measure serve as the basis for setting up capital requirement. Here, we explain why these properties are reasonable requirements in the context of capital requirements.

- (Translativity) This property means that if the loss of the business increases (decreases) by a fixed amount then the business should raise (lower) its capital by the same amount to account for the increased (decreased) risk. Since there is no added uncertainty, the amount of increase (decrease) in the capital should exactly match the amount of increase (decrease) in loss.
- (Positive homogeneity) This property means that if the loss is inflated by a factor c, then the capital should increase by the same factor.
- (Subadditivity) This property can be roughly translated as "diversification reduces risk". The capital required for the sum of two risks should be less than the sum

of standalone capital requirements for the two risks. The amount of reduction in capital by combining two risks is viewed as the benefit of "diversification".
- (Monotonicity) This property means that capital requirement should be commensurate with the size of the underlying risk. Larger risk requires higher capital.

If a decision maker is willing to accept coherence as a guiding principle of risk management, then there are a number of well-known coherent risk measures available for applications. Before showing examples of coherent risk measure, let us first consider the earlier-mentioned risk measures.

Example 2.7 In general, the Value-at-Risk is not coherent. To see this, we look at various properties of a coherent risk measure.

(a) *VaR is translative, positive homogeneous, and monotonic.* Observe that for any real number c and $p \in (0, 1)$,

$$\begin{aligned}\mathrm{VaR}_p[X+c] &= \inf\{x : \mathbb{P}(X+c \le x) \ge p\} = \inf\{x : \mathbb{P}(X \le x-c) \ge p\} \\ &= \inf\{y+c : \mathbb{P}(X \le y) \ge p\} = \inf\{y : \mathbb{P}(X \le y) \ge p\} + c \\ &= \mathrm{VaR}_p[X] + c.\end{aligned}$$

Similarly, one can show its homogeneity. For any constant $c > 0$,

$$\begin{aligned}\mathrm{VaR}_p[cX] &= \inf\{x : \mathbb{P}(cX \le x) \ge p\} = \inf\{x : \mathbb{P}(X \le x/c) \ge p\} \\ &= \inf\{cy : \mathbb{P}(X \le y) \ge p\} = c\inf\{y : \mathbb{P}(X \le y) \ge p\} = c\mathrm{VaR}_p[X].\end{aligned}$$

Suppose that $\mathbb{P}(X \le Y) = 1$. If $Y \le y$ for any fixed y, then it is obvious that $X \le y$. Therefore, the set $\{Y \le y\} \subseteq \{X \le y\}$, implying that $\mathbb{P}(Y \le y) \le \mathbb{P}(X \le y)$. Consequently, $\mathbb{P}(Y \le y) \ge p$ implies $\mathbb{P}(X \le y) \ge p$, which means that $\{y : \mathbb{P}(Y \le y) \ge p\} \subseteq \{y : \mathbb{P}(X \le y) \ge p\}$. Since the infimum of a smaller set is no smaller than that of a larger set, then

$$\mathrm{VaR}_p[Y] = \inf\{y : \mathbb{P}(Y \le y) \ge p\} \ge \inf\{y : \mathbb{P}(X \le y) \ge p\} = \mathrm{VaR}_p[X],$$

which proves its monotonicity.

(b) However, *VaR is not subadditive.* See Problem 2 for an illustration.

Example 2.8 Since TVaR_p is by definition an "arithmetic" average of VaR_q's at all levels between p and 1, TVaR inherits many properties of VaR. It can be shown that TVaR is translative, positively homogeneous, and monotonic. However, in contrast with VaR, TVaR is subadditive and hence a coherent risk measure.

To this end, we show that TVaR is subadditive. Recall the representation of TVaR as an optimization problem in (2.5). Choosing a constant $c = \mathrm{VaR}_p[X] + \mathrm{VaR}_p[Y]$, we must have

$$\begin{aligned}\mathrm{TVaR}_p[X+Y] &\le \frac{1}{1-p}\mathbb{E}[(X+Y-c)_+]+c \\ &\le \frac{1}{1-p}\mathbb{E}[(X-\mathrm{VaR}_p[X])_+ + (Y-\mathrm{VaR}_p[Y])_+]+c \\ &= \mathrm{TVaR}_p[X]+\mathrm{TVaR}_p[Y],\end{aligned}$$

where the second inequality comes from the elementary inequality $(a+b)_+ \le (a)_+ + (b)_+$ for any $a, b \in \mathbb{R}$. This completes the proof of its subadditivity.

2.3 Ordering of Risks

While the comparison of risk management plans is often done by qualitative means, we could also use some mathematical tools to compare the outcomes of risk management plans. One classical tool is stochastic ordering known in the probability theory to compare random variables. The purpose of stochastic ordering is to establish a relation between any pair of random variables by a certain criterion.

One could argue that we can already compare risks by the previously defined risk measures. For example, if we use the 90% quantile as the basis of setting up the capital to cover risks, then we can use the size of the required capital as a gauge of riskiness. Say, if $\mathrm{VaR}_{0.9}[X] \le \mathrm{VaR}_{0.9}[Y]$, then it costs more capital to cover the risk Y than it is to cover the risk X. One could say that Y is riskier than X. While this is essentially the basis of the ordering, such a comparison is a "local" property that focuses only on the 90% quantile and lacks consideration at a "global" level. What if it is more costly to pay for losses from the risk X at the 1% worst scenarios than it is to pay for the risk Y? Therefore, we need to compare the risks at different quantile levels.

2.3.1 *Stochastic Order*

Definition 2.6 Given two random variables X and Y, X is said to be less than or equal to Y in stochastic order, denoted by $X \le_{st} Y$ if

$$F_X(z) \ge F_Y(z), \tag{2.6}$$

for all $z \in (-\infty, \infty)$. Or equivalently, for all $p \in (0, 1)$,

$$\mathrm{VaR}_p[X] \le \mathrm{VaR}_p[Y]. \tag{2.7}$$

Example 2.9 Let X be the normal random variable with the mean μ_X and the variance σ_X^2 and Y be the normal random variable with the mean μ_Y and the variance σ_Y^2. Then, note that

$$X \le_{st} Y \quad \text{if and only if} \quad \mu_X \le \mu_Y,\ \sigma_X^2 = \sigma_Y^2.$$

Proof Consider the case that $X \le_{st} Y$. Assume that $\sigma_X^2 \neq \sigma_Y^2$. Say, $\sigma_X^2 \le \sigma_Y^2$. Note that

$$\text{VaR}_p[X] - \text{VaR}_p[Y] = \mu_X + \sigma_X \Phi^{-1}(p) - \mu_Y - \sigma_Y \Phi^{-1}(p) \to -\infty,$$

as $p \to 0$. It follows that there exists $p \in (0, 1)$ such that $\text{VaR}_p[X] \ge \text{VaR}_p[Y]$, which is a contradiction to the fact that $X \le_{st} Y$. Similar arguments apply for $\sigma_X^2 \ge \sigma_Y^2$. Thus, $\sigma_X^2 = \sigma_Y^2$. Assume that $\mu_X > \mu_Y$. Then,

$$\text{VaR}_p[X] - \text{VaR}_p[Y] = \mu_X + \sigma_X \Phi^{-1}(p) - \mu_Y - \sigma_Y \Phi^{-1}(p) = \mu_X - \mu_Y > 0,$$

which contradicts the fact that $X \le_{st} Y$. Therefore, $\mu_X \le \mu_Y$.

Now consider that $\mu_X \le \mu_Y$ and $\sigma_X^2 = \sigma_Y^2$. It follows immediately that

$$\text{VaR}_p[X] - \text{VaR}_p[Y] = \mu_X + \sigma_X \Phi^{-1}(p) - \mu_Y - \sigma_Y \Phi^{-1}(p) \le 0.$$

2.3.2 *Convex Order*

In many financial and insurance applications, the ordering of random variables plays a critical role in understanding the differences between comparable risks. We shall use the concept of convex ordering to demonstrate the effect of diversification later in this section.

Definition 2.7 Given two random variables X and Y, X is said to be less than or equal to Y in convex order, denoted by $X \le_{cx} Y$ if

$$\mathbb{E}[v(X)] \le \mathbb{E}[v(Y)], \tag{2.8}$$

for any convex function v such that the expectations exist.

Convex ordering is used to compare random variables with identical means. Note that $x \mapsto x$ and $x \mapsto -x$ are both convex functions. Then it follows from the definition that if $X \le_{cx} Y$ then $\mathbb{E}[X] \le \mathbb{E}[Y]$ and $\mathbb{E}[X] \ge \mathbb{E}[Y]$, which implies

$\mathbb{E}[X] = \mathbb{E}[Y]$. Convex order also implies the degree of dispersion for random variables. If $X \leq_{cx} Y$ then $\mathbb{V}[X] \leq \mathbb{V}[Y]$. One can prove this result by letting $v(x) = x^2$ in the definition and using the fact that $\mathbb{E}[X] = \mathbb{E}[Y]$. However, one should keep in mind that convex order is much stronger than the order of variances, because the inequality (2.8) holds for more than just linear and quadratic functions.

Example 2.10 Let X be a normal random variable with the mean μ_X and the variance σ_X^2 and Y be a normal random variable with the mean μ_Y and the variance σ_Y^2. From the discussion above, it naturally follows that

$$X \leq_{cx} Y \quad \text{if and only if} \quad \mu_x = \mu_Y,\ \sigma_X^2 \leq \sigma_Y^2.$$

Proof By the definition, it must hold that $\mathbb{E}[X] = \mathbb{E}[Y]$. Moreover, by using the convex function $v(x) = x^2$, it follows that

$$\begin{aligned} &\mathbb{E}[X^2] \leq \mathbb{E}[Y^2] \\ \iff &\mathbb{E}[X^2] - (\mathbb{E}[X])^2 \leq \mathbb{E}[Y^2] - (\mathbb{E}[Y])^2 \\ \iff &\mathbb{V}[X] \leq \mathbb{V}[Y]. \end{aligned}$$

Example 2.11 Let X and Y be two random variables with the order $X \leq_{cx} Y$ and the random variable Z is independent of both X and Y. Show that

$$X + Z \leq_{cx} Y + Z. \tag{2.9}$$

Proof We can prove the result by definition. Note that for any convex function v and any real number z, the function $x \mapsto v(x + z)$ must also be convex. Then $X \leq_{cx} Y$ implies that $\mathbb{E}[v(X + z)] \leq \mathbb{E}[v(Y + z)]$. Hence,

$$\begin{aligned} \mathbb{E}[v(X + Z)] &= \int_{-\infty}^{\infty} \mathbb{E}[v(X + z)]\mathbb{P}(Z \in \mathrm{d}z) \\ &\leq \int_{-\infty}^{\infty} \mathbb{E}[v(Y + z)]\mathbb{P}(Z \in \mathrm{d}z) = \mathbb{E}[v(Y + Z)], \end{aligned}$$

which implies (2.9) by definition.

It is in general difficult to check the convex order by attempting to verify (2.8) for all convex functions. Therefore, we often resort to a number of other equivalent or sufficient conditions for convex order.

Theorem 2.1 *Given two random variables X and Y, $X \leq_{cx} Y$ if and only if, for all $d \in R$,*

$$\mathbb{E}[(X-d)_+] \leq \mathbb{E}[(Y-d)_+], \qquad \mathbb{E}[(d-X)_+] \leq \mathbb{E}[(d-Y)_+]. \quad (2.10)$$

It is easy to see that (2.10) is a necessary condition for convex order, because $x \mapsto (x-d)_+$ and $x \mapsto (d-x)_+$ are both convex functions. One can also prove that (2.10) is sufficient because any convex function can be approximated as closely as desired by a linear combination of functions of the form $(x-d)_+$ and $(d-x)_+$.

It is often easy to verify a convex order using a simple sufficient condition, also known as the *Karlin-Novikoff cut-off criterion*, which states that distribution functions of X and Y cross each other exactly once.

Theorem 2.2 *Let F_X and F_Y be the distribution functions of random variables X and Y, respectively. If there exists $t^* \in \mathbb{R}$ for which*

$$\begin{cases} F_X(t) \leq F_Y(t), & t < t^*; \\ F_X(t) \geq F_Y(t), & t > t^*, \end{cases}$$

and $\mathbb{E}[X] = \mathbb{E}[Y]$, then $X \leq_{cx} Y$.

Proof Consider the difference

$$\Delta(d) := \mathbb{E}[(Y-d)_+] - \mathbb{E}[(X-d)_+]. \quad (2.11)$$

Using the identity (2.17), we obtain that

$$\Delta(d) = \int_d^\infty (F_X(t) - F_Y(t))\, dt = \int_d^\infty \left(\overline{F}_Y(t) - \overline{F}_X(t)\right) dt.$$

Note that we can rewrite

$$\Delta(d) = \left(\int_0^\infty \overline{F}_Y(t)\, dt - \int_d^0 F_Y(t)\, dt\right) - \left(\int_0^\infty \overline{F}_X(t)\, dt - \int_d^0 F_X(t)\, dt\right). \quad (2.12)$$

It follows from integration by parts that

$$\mathbb{E}[X] = \int_0^\infty \overline{F}_X(t)\, dt - \int_{-\infty}^0 F_X(t)\, dt.$$

Therefore, as we let $d \to -\infty$ in (2.12), it must be true that

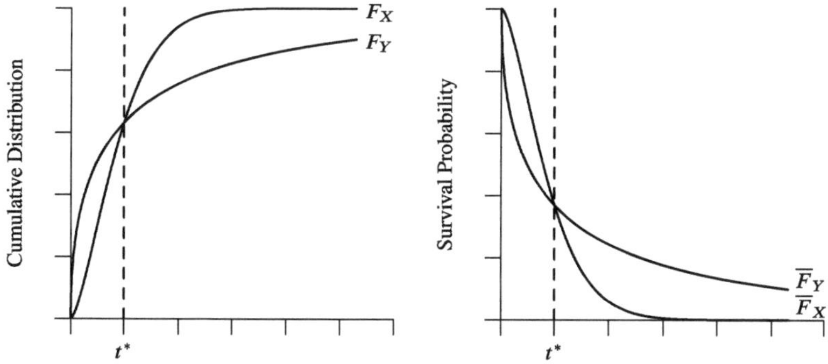

Fig. 2.4 Comparison on thickness of tail

$$\Delta(d) \to \mathbb{E}[Y] - \mathbb{E}[X] = 0.$$

It is also obvious from (2.11) that $\Delta(d) \to 0$ as $d \to \infty$. By assumption, we know that $\Delta'(d) = F_X(d) - F_Y(d) \geq 0$ for $d \leq t^*$ and $\Delta'(d) \leq 0$ for $d > t^*$. Therefore, the function Δ is non-decreasing over $(-\infty, t^*]$ and non-increasing over (t^*, ∞), which implies that $\Delta(d) \geq 0$ for all $d \in \mathbb{R}$. Using similar arguments, one can also show that for all $d \in \mathbb{R}$,

$$\nabla(d) := \mathbb{E}[(d - Y)_+] - \mathbb{E}[(d - X)_+] \geq 0.$$

Applying Theorem 2.1, we obtain $X \leq_{cx} Y$.

As shown in Fig. 2.4, the graph of the survival function $\overline{F}_Y$ lies above that of $\overline{F}_X$ to the right of t^*. In simple words, it is more likely to observe large values of Y than large values of X. Sometimes, this relationship is referred to in the literature as Y being *thicker-tailed* than X. From an insurer's point of view, a thicker-tailed risk is more costly than otherwise. For example, the cost of stop-loss reinsurance (see definition in Example 2.18) is typically estimated by $\mathbb{E}[(X - u)_+]$ where u is the policy limit of the primary insurance company. If Y has a thicker tail than X, then Theorem 2.2 implies that $\mathbb{E}[(X - u)_+] \leq \mathbb{E}[(Y - u)_+]$ for any $u \in \mathbb{R}$, which implies that a stop-loss insurance policy costs less on average to cover the risk X than it is to cover Y.

Example 2.12 Let $\mathcal{X}$ be a collection of all random variables with the same support $[a, b]$ where $0 < a < b < \infty$ and the same mean μ for $a < \mu < b$. Consider the distribution function of any arbitrary random variable in $\mathcal{X}$, whose graph is represented by the solid curve in Fig. 2.5. Recall that $\int_0^b \overline{F}(x)\,dx = \mathbb{E}[X]$. For any arbitrary random variable $X \in \mathcal{X}$, the area of the shaded region should be equal to μ. We can use the concept of thicker-tailedness in Theorem 2.2 to identify two extreme cases in convex order.

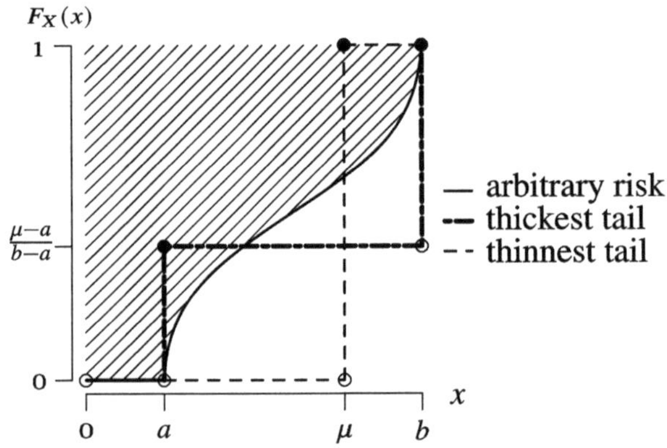

Fig. 2.5 Random variables with the same finite support and mean

(Random variable with the thinnest tail) We look for such a random variable X_* that

$$X_* \leq_{cx} X, \qquad \text{for any } X \in \mathcal{X}.$$

It is clear from Fig. 2.4 that the distribution function of X_* must lie below that of any arbitrary distribution function near a. Then it is obvious that such a distribution must be 0 on $[a, t^*]$ for the largest possible t^*. In such a case, the distribution of X_* must jump from 0 to 1 at t^*. Keep in mind that the area above its distribution function and the horizontal line $F(x) = 1$ must be equal to μ. Then the total area is given by $t^* = \mu$. In other words, the random variable is in fact a constant,

$$X_* = \mu \qquad \text{with probability 1.}$$

The graph of the distribution function of X_* is shown in Fig. 2.5.

(Random variable with the thickest tail) We turn to the other extreme case where

$$X \leq_{cx} X^*, \qquad \text{for all } X \in \mathcal{X}.$$

It is also clear from Fig. 2.4 that the distribution function X^* must lie above that of any arbitrary distribution near a. It means that such a distribution function must be some constant c on $[a, b]$ for the largest possible c. Hence, the distribution function of X^* should have a jump from 0 to some constant c at a and again another jump from c to 1 at b. The total area between such a distribution function and the top horizontal line must be $a + (1 - c)(b - a)$. Since all random variables have the same mean, we must have $c = (b - \mu)/(b - a)$. In other words, the random variable X^* can be represented by

$$X^* = \begin{cases} a, & \text{with probability } \frac{b-\mu}{b-a}; \\ b, & \text{with probability } \frac{\mu-a}{b-a}. \end{cases}$$

The graph of the distribution function X^* is also shown in Fig. 2.5. According to Theorem 2.2, for any arbitrary random variable $X \in \mathcal{X}$,

$$X_* \leq_{cx} X \leq_{cx} X^*.$$

Another important sufficient condition for a convex order is that a random variable is always no less than its conditional expectation in convex order.

Theorem 2.3 *If $X = \mathbb{E}[Y|Z]$ for some random variable Z, then*

$$X \leq_{cx} Y.$$

Proof This is a straightforward application of Jensen's inequality in Appendix 2.C. For any arbitrary convex function u,

$$\mathbb{E}[u(X)] = \mathbb{E}\left[u\big(\mathbb{E}[Y|Z]\big)\right] \leq \mathbb{E}\left[\mathbb{E}[u(Y)|Z]\right] = \mathbb{E}[u(Y)],$$

provided that the integrals exist. By definition, $X \leq_{cx} Y$.

Example 2.13 Let X and Y be two independent random variables with $\mathbb{E}[X] < \infty$ and $\mathbb{E}[Y] = 0$. Show that

$$X \leq_{cx} X + Y.$$

Proof Observe that $X = \mathbb{E}[X + Y|X] = X + \mathbb{E}[Y]$. It follows immediately from Theorem 2.3 that $X \leq_{cx} X + Y$.

The importance of convex order in the understanding of diversification lies in the fact that tail risk can be measured by risk measures such as TVaR and that convex order and its relationship with TVaR can be used to quantify the effect of diversification.

Theorem 2.4 *Given two random variables X and Y, if $X \leq_{cx} Y$ then for all $p \in (0, 1)$,*

$$\mathrm{TVaR}_p[X] \leq \mathrm{TVaR}_p(Y).$$

Proof Recall the representation of TVaR in (2.5). Then

$$\begin{aligned}\mathrm{TVaR}_p[X] &\leq \mathrm{VaR}_p(Y) + \frac{1}{1-p}\mathbb{E}[(X - \mathrm{VaR}_p[Y])_+] \\ &\leq \mathrm{VaR}_p(Y) + \frac{1}{1-p}\mathbb{E}[(Y - \mathrm{VaR}_p[Y])_+] = \mathrm{TVaR}_p[Y],\end{aligned}$$

where the second inequality is due to the fact that $x \mapsto (x - \mathrm{VaR}_p[X])_+$ is a convex function.

Example 2.14 Let X and Y be two normally distributed random variables with the mean μ_X, μ_Y and the variance σ_X^2, σ_Y^2, respectively. Suppose that $\mu_X = \mu_Y$ and $\sigma_X^2 \leq \sigma_Y^2$. It is known from Example 2.10 that $X \leq_{cx} Y$. Observe that, for all $p \in (0, 1)$,

$$\begin{aligned}\mathrm{TVaR}_p[X] - \mathrm{TVaR}_p[Y] &= \mu_X + \sigma_X \frac{\phi(\Phi^{-1}(p))}{1-p} - \mu_Y + \sigma_Y \frac{\phi(\Phi^{-1}(p))}{1-p} \\ &= (\sigma_X - \sigma_Y)\frac{\phi(\Phi^{-1}(p))}{1-p} \leq 0.\end{aligned}$$

This calculation confirms the result from Theorem 2.4 for the case of normal random variables.

Readers who are interested in more details of convex order are referred to Shaked and Shanthikumar (1994).

2.4 Multivariate Modeling

Risks are often interconnected. For example, climate change affects water supply and agricultural production, which leads to extreme volatilities in agricultural commodity prices. Limited agricultural production can also contribute to geopolitical risks, such as economic collapse and social unrest. If we want to assess the risk of investment involving commodity prices, it is sometimes also necessary to understand other factors like climate change and economic fundamentals such as supply chains and income inequalities. In this section, we consider quantitative models to understand and assess the interrelationships of dependent risks.

2.4.1 Multivariate Distribution

Consider a d-dimensional random vector of (interconnected) risks $\boldsymbol{X} = (X_1, \ldots, X_d)^\top$. Its joint distribution function is given by

$$F_X(\boldsymbol{x}) = \mathbb{P}(\boldsymbol{X} \leq \boldsymbol{x}) = \mathbb{P}(X_1 \leq x_1, ..., X_d \leq x_d),$$

where $\boldsymbol{x} = (x_1, \cdots, x_d)^\top$ is a d-dimensional vector of fixed numbers. The random vector $\boldsymbol{X}$ is absolutely continuous if

$$F_X(\boldsymbol{x}) = \int_{-\infty}^{x_1} ... \int_{-\infty}^{x_d} f_X(u_1, ..., u_d) du_1 ... du_d$$

for some non-negative function f_X, which is called the joint density function. We omit the subscript $\boldsymbol{X}$ when no ambiguity arises. A marginal distribution is the distribution function of some but not all components of $\boldsymbol{X}$. For example, let $\widetilde{\boldsymbol{X}} = (X_{i1}, X_{i2}, ..., X_{ik})^\top$ for integers $k < d$, $1 \leq i_1, ..., i_k \leq d$.

$$\begin{aligned} F_{\widetilde{X}}(\widetilde{\boldsymbol{x}}) &= \mathbb{P}(\widetilde{\boldsymbol{X}} \leq \widetilde{\boldsymbol{x}}) = \mathbb{P}(X_{i1} \leq x_1, ..., X_{ik} \leq x_k) \\ &= \lim_{x_j \to \infty, j \neq i_1, ..., i_k} F_X(\boldsymbol{x}) \end{aligned}$$

for some $\widetilde{\boldsymbol{x}} = (x, ..., x_k)^\top$. If only a single component is considered, i.e. $\widetilde{\boldsymbol{X}} = (X_i)$, then we often write $F_i(x_i)$ for brevity. The joint survival function of the random vector $\boldsymbol{X}$ is given by

$$\overline{F}_X(\boldsymbol{x}) = \mathbb{P}(\boldsymbol{X} > \boldsymbol{x}) = \mathbb{P}(X_1 > x_1, ..., X_d > x_d).$$

The marginal survival function of components of $\boldsymbol{X}$ is given by

$$\begin{aligned} \overline{F}_{\widetilde{X}}(\widetilde{\boldsymbol{x}}) &= \mathbb{P}(\widetilde{\boldsymbol{X}} > \widetilde{\boldsymbol{x}}) = \mathbb{P}(X_{i1} > x_1, ..., X_{ik} > x_k) \\ &= \lim_{\substack{x_j \to -\infty \\ j \neq i_1, ..., i_k \\ i \leq j \leq d}} \overline{F}_X(\boldsymbol{x}). \end{aligned}$$

For any random couple (X, Y) with finite variances, its covariance is defined by

$$\begin{aligned} \mathbb{C}[X, Y] &= \mathbb{E}[(X - \mathbb{E}[X])(Y - \mathbb{E}[Y])] \\ &= \int_{-\infty}^{\infty} \int_{-\infty}^{\infty} (x - \mathbb{E}[X])(y - \mathbb{E}[Y]) f_{X,Y}(x, y) \, \mathrm{d}x \, \mathrm{d}y, \end{aligned}$$

provided that such an integral exists. Observe that the variance of any random variable X can be obtained from $\mathbb{V}[X] = \mathbb{C}[X, X]$.

2.4.2 Conditional Distribution

Suppose there are two absolutely continuous random vectors $\boldsymbol{X}_1$ and $\boldsymbol{X}_2$ under consideration. We shall write the combined vector $\boldsymbol{X} = (\boldsymbol{X}_1, \boldsymbol{X}_2)$. The conditional distribution of $\boldsymbol{X}_2$ given $\boldsymbol{X}_1 = \boldsymbol{x}_1$ has the conditional density

$$f_{\boldsymbol{X}_2|\boldsymbol{X}_1}(\boldsymbol{x}_2|\boldsymbol{x}_1) = \frac{f_{\boldsymbol{X}}(\boldsymbol{x}_1, \boldsymbol{x}_2)}{f_{\boldsymbol{X}_1}(\boldsymbol{x}_1)}. \tag{2.13}$$

Two random vectors are said to be independent if

$$F_{\boldsymbol{X}}(\boldsymbol{x}) = F_{\boldsymbol{X}_1}(\boldsymbol{x}_1)F_{\boldsymbol{X}_2}(\boldsymbol{x}_2)$$

for all $(\boldsymbol{x}_1, \boldsymbol{x}_2)$ on its domain. If both vectors are absolutely continuous then

$$f_{\boldsymbol{X}}(\boldsymbol{x}) = f_{\boldsymbol{X}_1}(\boldsymbol{x}_1)f_{\boldsymbol{X}_2}(\boldsymbol{x}_2). \tag{2.14}$$

Inserting (2.14) into (2.13) shows that

$$f_{\boldsymbol{X}_2|\boldsymbol{X}_1}(\boldsymbol{x}_2|\boldsymbol{x}_1) = f_{\boldsymbol{X}_2}(\boldsymbol{x}_2)$$

which implies that the conditioning of $\boldsymbol{X}_1$ has no effect on the probability distribution of $\boldsymbol{X}_2$. This result agrees with the interpretation of independence in layman's terms.

Example 2.15 Consider a continuous random vector (X, Y) whose joint density function is given by

$$f_{X,Y}(x, y) = \frac{1}{b-a} y e^{-xy}, \qquad x > 0, \;\; 0 < a \le y \le b.$$

Note that the marginal distribution function of Y can be obtained from

$$\begin{aligned} F_Y(y) = \lim_{x\to\infty} F_{X,Y}(x, y) &= \lim_{x\to\infty} \int_{-\infty}^{x} \int_{-\infty}^{y} f_{X,Y}(u, v)\, \mathrm{d}u\mathrm{d}v \\ &= \int_{-\infty}^{y} \left(\int_{-\infty}^{\infty} f_{X,Y}(u, v)\, \mathrm{d}u \right) \mathrm{d}v. \end{aligned}$$

It implies that the marginal density function of Y can be determined by

$$f_Y(y) = \int_{-\infty}^{\infty} f_{X,Y}(u, y)\, \mathrm{d}u = \frac{1}{b-a}, \qquad a \le x \le b.$$

Then the conditional distribution of X given $Y = y$ can be found by

$$f_{X|Y}(x|y) = \frac{f_{X,Y}(x, y)}{f_Y(y)} = y e^{-xy}, \qquad x > 0.$$

We shall use the concept of conditional mean as a way of allocating risks in later chapters. The conditional mean of a random variable, say X, given the value of another random variable, say $Y = y$, is defined by

$$\mathbb{E}[X|Y=y]=\int_{-\infty}^{\infty} x f_{X|Y}(x|y)\,\mathrm{d}x.$$

Observe that the conditional mean of one random variable given the value of another random variable is a real number and can be viewed as a real-valued function, i.e. $\mathbb{E}[X|Y=y]=g(y)$ where y is treated as an argument or a parameter. We sometimes consider a more general conditional mean without specifying the value of the conditioning random variable. Then the conditional mean is no longer a number, but rather a random variable itself:

$$\mathbb{E}[X|Y]=g(Y).$$

Note that this conditional mean random variable has the same mean as the random variable without any conditioning, i.e. $\mathbb{E}[\mathbb{E}[X|Y]]=\mathbb{E}[g(Y)]=\mathbb{E}[X]$, due to the fact that

$$\begin{aligned}\mathbb{E}[g(Y)] &= \int_{-\infty}^{\infty} g(y) f_Y(y)\mathrm{d}y = \int_{-\infty}^{\infty}\left(\int_{-\infty}^{\infty} x f_{X|Y}(x|y)\,\mathrm{d}x\right) f_Y(y)\mathrm{d}y \\ &= \int_{-\infty}^{\infty} x\left(\int_{-\infty}^{\infty} f_{X,Y}(x,y)\,\mathrm{d}y\right)\mathrm{d}x = \int_{-\infty}^{\infty} x f_X(x)\,\mathrm{d}x = \mathbb{E}[X].\end{aligned}$$

Example 2.16 Continue with the previous example. Observe that, for $a \le y \le b$,

$$\mathbb{E}[X|Y=y]=\int_0^{\infty} xye^{-xy}\,\mathrm{d}x = \frac{1}{y}.$$

In general, the conditional mean of X given Y is given by

$$\mathbb{E}[X|Y]=\frac{1}{Y}.$$

2.4.3 *Multivariate Normal Distribution*

One of the most commonly used multivariate distributions is the normal distribution. A normal random vector can be constructed from a set of independent standard normal random variables.

Definition 2.8 $X = (X_1, ..., X_d)^\top$ has a multivariate normal distribution if

$$X = \mu + AZ,$$

where $Z = (Z_1, ..., Z_r)$ is a vector of independent standard normal random variables, $\mu \in \mathbb{R}^d$ is a vector of fixed numbers, and $A \in \mathbb{R}^{d\times r}$ is a matrix of fixed numbers.

The normal random vector is absolutely continuous and has the following joint density function

$$f(x) = \frac{1}{(2\pi)^{\frac{d}{2}}|\Sigma|^{\frac{1}{2}}} \exp\left\{-\frac{1}{2}(x-\mu)^\top \Sigma^{-1}(x-\mu)\right\},$$

with some constant vector μ and positive semi-definite matrix Σ for any $x \in \mathbb{R}^d$. Note that the density depends on x only through the quadratic form $(x-\mu)^\top\Sigma^{-1}(x-\mu)$. Therefore, if we draw a contour plot of the density function, i.e. curves on which the density function has a fixed value, the curves would become ellipsoids. See Fig. 2.6. Each ellipsoid represents the set of points $\{x \in \mathbb{R}^d | (x-\mu)^\top \Sigma^{-1}(x-\mu) = c\}$ for some given constant c.

Observe that the mean of a multivariate normal distribution is given by

$$E[X] = \mu + AE[Z] = \mu.$$

The covariance matrix is given by

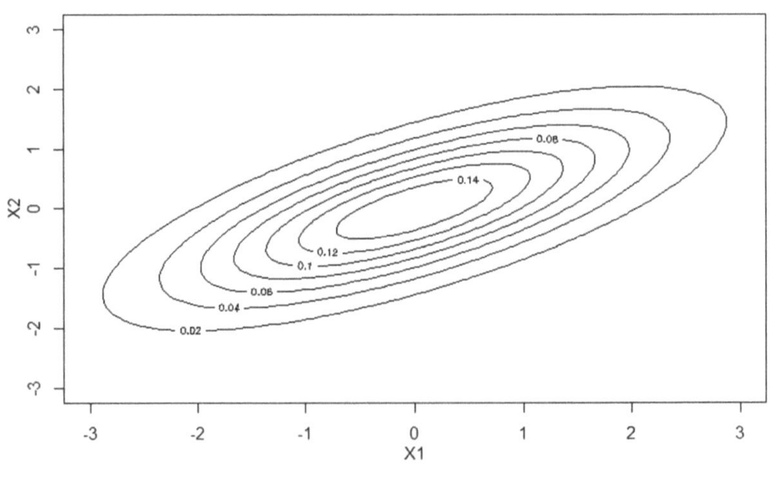

Fig. 2.6 Contour plot for bivariate normal distribution

$$\mathbb{C}[\boldsymbol{X}] = \boldsymbol{A}\mathbb{V}[\boldsymbol{Z}]\boldsymbol{A}^\top = \boldsymbol{A}\boldsymbol{I}\boldsymbol{A}^\top = \boldsymbol{A}\boldsymbol{A}^\top = \boldsymbol{\Sigma}.$$

We sometimes write $\boldsymbol{X} \sim N(\boldsymbol{\mu}, \boldsymbol{\Sigma})$ for brevity.

An important fact about multivariate normal random vector is that, if their components have pairwise zero covariance, then their components are independent of each other. It is easy to prove this by showing that, when $\boldsymbol{\Sigma} = \boldsymbol{I}$,

$$f(\boldsymbol{x}) = f_1(x_1)f_2(x_2)\cdots f_d(x_d),$$

where f_i is the marginal density function of the normally distributed component X_i.

The conditional means of normal random variables are useful in the context of risk sharing plans in later chapters.

Example 2.17 (*Conditional Mean of Normal Distribution*) Suppose that two random variables (X_1, X_2) are normally distributed. Show that

$$\mathbb{E}[X_1|X_2] = \mathbb{E}[X_1] + \frac{\mathbb{C}[X_1, X_2]}{\mathbb{V}[X_2]}(X_2 - \mathbb{E}[X_2]).$$

Proof Suppose that X_1 and X_2 have non-zero covariance. Define

$$\hat{X} = \frac{\mathbb{C}[X_1, X_2]}{\mathbb{V}[X_2]}X_2, \quad \tilde{X} = X_1 - \hat{X}.$$

Observe that

$$\mathbb{C}[X_2, \tilde{X}] = \mathbb{C}[X_2, X_1] - \mathbb{C}[X_2, \hat{X}] = 0,$$

which implies that X_2 and $\tilde{X}$ are uncorrelated and hence independent given that both are normal. Note that $X_1 = \tilde{X} + \hat{X}$. By taking conditional expectations on both sides given X_2, we obtain that

$$\begin{aligned}\mathbb{E}[X_1|X_2] &= \mathbb{C}[X_2, X_1]\mathbb{E}[X_2|X_2] + \mathbb{E}[\tilde{X}|X_2]\\ &= \mathbb{C}[X_2, X_1]X_2 + \mathbb{E}[\tilde{X}]\\ &= \mathbb{E}[X_1] + \frac{\mathbb{C}[X_1, X_2]}{\mathbb{V}[X_2]}(X_2 - \mathbb{E}[X_2]).\end{aligned}$$

2.4.4 Comonotonicity

It is often the case that we want to understand the impact of extreme economic scenarios. These extreme cases are sometimes related to particular dependence relationships between random variables. For example, during an economic recession,

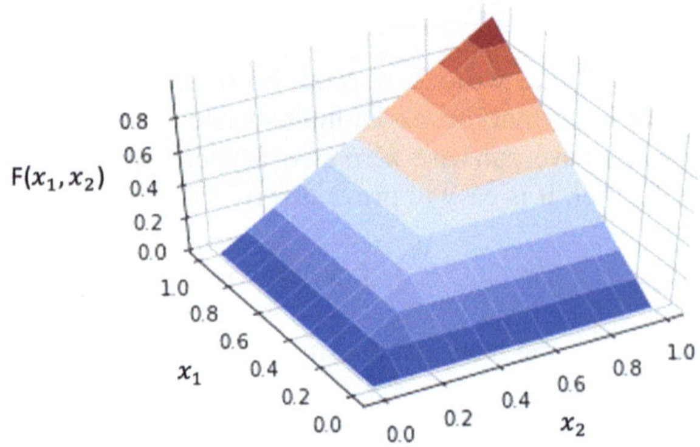

Fig. 2.7 Joint distribution function of comonotonic uniform random variables

risk factors such as stock prices and housing prices may all go down together. In this section, we discuss two extreme cases of dependence between two random variables, one case in which the two move in the same direction and the other case in which the two move in the opposite direction.

Suppose that a pair of random variables X_1 and X_2 is known to have marginal distributions F_1 and F_2 and the joint distribution F. It can be shown that any distribution function F is bounded by

$$(F_1(x) + F_2(y) - 1)_+ \leq F(x, y) \leq F_1(x) \wedge F_2(y), \tag{2.15}$$

where the symbol $\wedge$ means the minimum, i.e. $x \wedge y = \min\{x, y\}$. We shall illustrate that the two bounds, also known as Fréchet lower and upper bounds, are in fact distribution functions themselves. The distribution function on the right side corresponds a comonotonic pair, whereas the distribution function on the left side represents a counter-monotonic pair.

Definition 2.9 A random vector $(X_1, X_2, \cdots, X_d)$ is said to be comonotonic if there exist a random variable Z and non-decreasing functions $t_1, t_2, \cdots, t_d$ such that $(t_1(Z), t_2(Z), \cdots, t_d(Z))$ has the same joint distribution as $(X_1, X_2, \cdots, X_d)$.

The intuition behind the comonotonicity is that these random variables tend to be small or large all together and there is a single "force" that drives the values of all random variables. To visualize these two cases, we can consider a pair of uniform random variables between 0 and 1. Figure 2.7 shows the joint distribution $F_1(x) \wedge F_2(y) = x \wedge y$ of two comonotonic uniform random variables for $0 \leq x, y \leq 1$.

There are many examples in the context of insurance that can be described by comonotonic random variables.

Example 2.18 Stop-loss insurance is a policy that limits claim coverage to a specified maximum amount. For instance, if the policy limit of stop-loss insurance is u and the total amount of insurance claims to the primary insurer is represented by X, then the reinsurer reimburses the primary insurer the amount by which the aggregate claim exceeds the policy limit, i.e.

$$X_2 = (X - u)_+,$$

and hence the primary insurer only suffers the total loss of

$$X_1 = X - X_2 = \min\{X, u\}.$$

In this case, the pair of losses for a primary insurer and reinsurer, (X_1, X_2), is clearly comonotonic.

Example 2.19 Another common form is proportional insurance, sometimes referred to as coinsurance or pro rata insurance, where the policyholder and the insurer share the loss under coverage proportionally. Suppose that the total loss from the policyholder is represented by X and the coinsurance rate is $\alpha \in (0, 1)$, then the policyholder pays a portion of the loss, i.e.

$$X_1 = \alpha X,$$

and the insurer pays the rest of the loss, i.e.

$$X_2 = X - X_1 = (1 - \alpha)X.$$

Hence, in this case, the total losses for the policyholder and the insurer (X_1, X_2) are also comonotonic.

In general, if we know the marginal distributions of (X_1, X_2), we can create a comonotonic pair of random variables (X_1^c, X_2^c) with the same marginal distribution as (X_1, X_2). Consider U to be uniform on $[0, 1]$. Then we can set $X_1^c = F_1^{-1}(U)$ and $X_2^c = F_2^{-1}(U)$. It follows that

$$\begin{aligned}\mathbb{P}(X_1^c \leq x, X_2^c \leq y) = \mathbb{P}\big(F_1^{-1}(U) \leq x, F_2^{-1}(U)\big) \leq y = \mathbb{P}(U \leq F_1(x), U \leq F_2(y))\\ = \mathbb{P}(U \leq F_1(x) \wedge F_2(y)) = F_1(x) \wedge F_2(y),\end{aligned}$$

which is precisely the upper bound in (2.15).

Example 2.20 Suppose that X_1, X_2 are two exponential random variables with mean λ_1, λ_2 respectively and with unknown dependence. Recall that the quantile function is given by $F_i^{-1}(p) = -\frac{1}{\lambda_i}\ln(1-p)$ for $i = 1, 2$. The comonotonic version of these two random variables can be given by

$$(X_1^c, X_2^c) = \left(-\frac{1}{\lambda_1}\ln U, -\frac{1}{\lambda_2}\ln U\right),$$

where U is a uniform random variable on [0, 1].

Theorem 2.5 (Additivity of TVaR for sums of comonotonic risks) *Consider a comonotonic random vector* $(X_1, X_2, ..., X_n)$*, and let* $S = X_1 + X_2 + \cdots + X_n$*. Then, for all* $p \in (0, 1)$*, we have*

$$\mathrm{TVaR}_p[S] = \sum_{i=1}^{n} \mathrm{TVaR}_p[X_i].$$

Proof Suppose that

$$\mathrm{VaR}_p[S] = \sum_{i=1}^{n} \mathrm{VaR}_p[X_i]. \tag{2.16}$$

Then, the desired result follows from (2.16) and the definition of TVaR. Thus, we only have to prove (2.16). Note we have $S = F_{X_1}^{-1}(U) + F_{X_2}^{-1}(U) + \cdots + F_{X_n}^{-1}(U) = g(U)$ with g a non-decreasing and left-continuous function. Thus,

$$F_S^{-1}(p) = F_{g(U)}^{-1}(p) = g(F_U^{-1}(p)) = g(p), \qquad 0 < p < 1,$$

and this implies the desired result.

2.4.5 *Counter-Monotonicity*

Definition 2.10 A bivariate random vector $\boldsymbol{X}$ is called counter-monotonic if $\boldsymbol{X}$ has the same distributions as $(t_1(Z), t_2(Z))$ for some random variable Z, where t_1 is a non-decreasing function and t_2 is a non-increasing function.

When we observe realizations of counter-monotonic random variables, it is often the case that one is small and the other is big. Or in other words, the two random

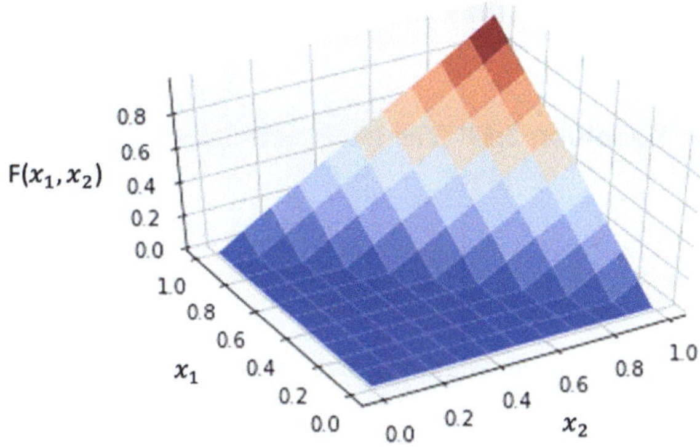

Fig. 2.8 Joint distribution function of counter-monotonic uniform random variables

variables tend to move in opposite directions. Figure 2.8 shows the joint distribution $(F_1(x) + F_2(y) - 1)_+ = (x + y - 1)_+$ of two counter-comonotonic uniform random variables.

In the bivariate case, counter-monotonic random variables with known marginal distributions F_1 and F_2 can be represented by $(F_1^{-1}(U), F_2^{-1}(1 - U))$, where U is a uniform random variable. Observe that

$$\begin{aligned}&\mathbb{P}(F_1^{-1}(U) \leq x, F_2^{-1}(1 - U) \leq y) = \mathbb{P}(U \leq F_1(x), 1 - U \leq F_2(y)) \\ &= \mathbb{P}(1 - F_2(x) \leq U \leq F_1(x), F_1(x) \geq 1 - F_2(y)) = (F_1(x) + F_2(y) - 1)_+ ,\end{aligned}$$

which is the lower bound in (2.15).

2.5 Dependence Measures

In earlier sections, we introduce a risk measure as summary statistics for key information of a univariate random variable, i.e.

$$X \to \rho(X),$$

where ρ is a mapping from a single random variable to a real number. We shall discuss a dependence measure as summary statistics for the interdependence of a pair of random variables,

$$(X, Y) \to \rho[X, Y],$$

where ρ is a mapping from a pair of random variables to a real number. The most used dependence measure is the Pearson correlation coefficient.

2.5.1 Pearson Correlation Coefficient

Definition 2.11 The linear correlation coefficient is given by

$$\rho[X, Y] = \frac{\mathbb{C}[X, Y]}{\sqrt{\mathbb{V}[X]\mathbb{V}[Y]}}.$$

Note that the Pearson correlation coefficient has the following property:

$$-1 \le \rho[X, Y] \le 1.$$

Example 2.21 It is easy to show that, if (X, Y) is a bivariate normal random vector with parameters

$$\mathbf{\Sigma} = \begin{pmatrix} \sigma_1^2 & \rho\sigma_1\sigma_2 \\ \rho\sigma_1\sigma_2 & \sigma_2^2 \end{pmatrix},$$

then

$$\mathbb{C}[X, Y] = \rho\sigma_1\sigma_2.$$

It follows by definition that

$$\rho[X, Y] = \rho.$$

Example 2.22 (*Perfect Correlation*) There are two cases of special interest for the correlation coefficient. For $a > 0$, $b \in \mathbb{R}$, suppose X and Y have the following linear transformation:

$$Y = aX + b.$$

Then the Pearson correlation coefficient is

$$\rho[X, Y] = \frac{\mathbb{C}[X, Y]}{\sqrt{\mathbb{V}[X]\mathbb{V}[Y]}} = 1,$$

where

$$\begin{aligned}
\mathbb{C}[X, aX] &= \mathbb{E}[aX^2] - \mathbb{E}[X]\mathbb{E}[aX] = a\mathbb{E}[X^2] - a(\mathbb{E}[X])^2 = a\mathbb{V}[X] \\
\mathbb{C}[X, b] &= \mathbb{E}[Xb] - \mathbb{E}[X]\mathbb{E}[b] = b\mathbb{E}[X] - b\mathbb{E}[X] = 0 \\
\mathbb{C}[X, Y] &= \mathbb{C}[X, aX + b] = a\mathbb{V}[X] \\
\mathbb{V}[Y] &= \mathbb{C}[Y, Y] = \mathbb{C}[aX + b, aX + b] = a^2\mathbb{C}[X, X] = a^2\mathbb{V}[X].
\end{aligned}$$

Similarly, one can show that for $a < 0$, $\rho[X, Y] = -1$.

Example 2.23 (*Imperfect Correlation*) The Pearson correlation coefficient is often referred to as a linear correlation coefficient because it is used to describe the tendency of linearity between two random variables. This is best illustrated through the following minimization problem

$$\min_{a,b\in\mathbb{R}} \mathbb{E}[(Y-(aX+b))^2],$$

where the optimal values of a^* and b^* are determined by

$$\begin{aligned} a^* &= \frac{\mathbb{C}[X,Y]}{\mathbb{V}[X]}, \\ b^* &= \mathbb{E}[Y]-a^*\mathbb{E}[X]. \end{aligned}$$

For any pair of random variables (X, Y) with given marginal distributions, the Pearson coefficient determines the slope of a linear function of X that is "closest" to Y. The closeness is measured by the mean squared error between Y and its predictor $aX+b$. We leave the proof as an exercise in Problem 7.

It should be pointed out that the Pearson correlation coefficient is a measure of linearity. Even if two random variables have a very small Pearson correlation coefficient, it does not necessarily mean that the two are close to independence or some "weak" dependence. The following example shows a case of "strong" dependence which has a zero correlation coefficient.

Example 2.24 If N is a standard normal random variable, the covariance between N and N^2 would be

$$\begin{aligned} \mathbb{C}[N,N^2] &= \mathbb{E}[NN^2]-\mathbb{E}[N]\mathbb{E}[N^2] \\ &= \mathbb{E}[N^3]-\mathbb{E}[N]\mathbb{E}[N^2] \\ &= 0. \end{aligned}$$

Then the Pearson correlation coefficient between them is

$$\rho[N,N^2]=0.$$

It is clear that the two random variables (N, N^2) are not independent and yet they have a zero correlation coefficient. This is owing to the fact that their relationship is quadratic, not linear.

It is also important to note that, while the Pearson correlation coefficient is bounded by ±1, it may not be able to reach them. The actual bounds of the coefficient are in fact determined by the marginal distributions of the two random variables.

Theorem 2.6 *If F_1 and F_2 are given marginal distributions of positive random variables X and Y, then the Pearson correlation coefficient between X and Y satisfies the following inequalities:*

$$\underline{\rho}(X, Y) \leq \rho[X, Y] \leq \overline{\rho}(X, Y),$$

where

$$\begin{aligned}\overline{\rho}(X, Y) &= \frac{\mathbb{C}[F_1^{-1}(U), F_2^{-1}(U)]}{\sqrt{\mathbb{V}[X]\mathbb{V}[Y]}},\\ \underline{\rho}(X, Y) &= \frac{\mathbb{C}[F_1^{-1}(U), F_2^{-1}(1-U)]}{\sqrt{\mathbb{V}[X]\mathbb{V}[Y]}}.\end{aligned}$$

Proof Firstly, the expectation of the cross term XY is

$$\begin{aligned}\mathbb{E}[XY] &= \int_0^\infty \int_0^\infty y_1 y_2 dF_X(y_1, y_2)\\ &= \int_0^\infty \int_0^\infty \int_0^{y_1} \int_0^{y_2} dY dX dF_X(y_1, y_2)\\ &= \int_0^\infty \int_0^\infty \int_u^\infty \int_v^\infty dF_X(y_1, y_2) dv du\\ &= \int_0^\infty \int_0^\infty \bar{F}_X(u, v) dv du.\end{aligned}$$

Recall the Fréchet bounds in (2.15) and that

$$\bar{F}_X(\boldsymbol{x}) = 1 - F_1(x_1) - F_2(x_2) + F_X(\boldsymbol{x}).$$

We know that

$$\begin{aligned}\mathbb{E}[F_1^{-1}(U)F_2^{-1}(1-U)] &\leq \mathbb{E}[XY] \leq \mathbb{E}[F_1^{-1}(U)F_2^{-1}(U)],\\ \mathbb{C}[F_1^{-1}(U), F_2^{-1}(1-U)] &\leq \mathbb{C}[X, Y] \leq \mathbb{C}[F_1^{-1}(U), F_2^{-1}(U)].\end{aligned}$$

Therefore, it can be shown that

$$\frac{\mathbb{C}[F_1^{-1}(U), F_2^{-1}(1-U)]}{\sqrt{\sigma^2(X)\sigma^2(Y)}} \leq \rho[X, Y] \leq \frac{\mathbb{C}[F_1^{-1}(U), F_2^{-1}(U)]}{\sqrt{\sigma^2(X)\sigma^2(Y)}}.$$

In the following example, we show that the range of the Pearson correlation coefficient between two log-normal random variables can be as narrow as possible in the neighborhood of zero.

Example 2.25 Suppose there are two log-normal random variables, i.e. for $i = 1, 2$,

$$X_i = e^{\mu_i + \sigma_i N_i},$$

where N_i is a standard normal random variable. To determine the left bound and right bound of the Pearson correlation coefficient in the proposition above, we have

$$\begin{aligned}
\mathbb{E}[X_i] &= e^{\mu_i + \frac{1}{2}\sigma_i^2}, \\
F_i^{-1}(U) &= e^{\mu_i + \sigma_i \Phi^{-1}(U)}, \\
\mathbb{V}[X_i] &= \mathbb{E}[X_i^2] - (\mathbb{E}[X_i])^2 = e^{2\mu_i + \sigma_i^2}(e^{\sigma_i^2} - 1), \\
\sqrt{\mathbb{V}[X_i]} &= e^{\mu_i + \frac{\sigma_i^2}{2}} \sqrt{e^{\sigma_i^2} - 1}.
\end{aligned}$$

Therefore, the covariance between $F_1^{-1}(U)$ and $F_2^{-1}(U)$ is

$$\begin{aligned}
\mathbb{C}[F_1^{-1}(U), F_2^{-1}(U)] &= \mathbb{E}[F_1^{-1}(U) F_2^{-1}(U)] - \mathbb{E}[F_1^{-1}(U)]\mathbb{E}[F_2^{-1}(U)] \\
&= \mathbb{E}[e^{\mu_1 + \mu_2 + (\sigma_1 + \sigma_2)\Phi^{-1}(U)}] - \mathbb{E}[X]\mathbb{E}[Y] \\
&= e^{\mu_1 + \mu_2 + \frac{1}{2}(\sigma_1 + \sigma_2)^2} - e^{\mu_1 + \mu_2 + \frac{1}{2}(\sigma_1^2 + \sigma_2^2)} \\
&= e^{\mu_1 + \mu_2 + \frac{\sigma_1^2 + \sigma_2^2}{2}}(e^{\sigma_1 \sigma_2} - 1).
\end{aligned}$$

Then the right bound of the proposition above is given by

$$\overline{\rho}[X_1, X_2] = \frac{\mathbb{C}[F_1^{-1}(U), F_2^{-1}(U)]}{\sqrt{\mathbb{V}[X_1]\mathbb{V}[X_2]}} = \frac{e^{\sigma_1 \sigma_2} - 1}{\sqrt{e^{\sigma_1^2} - 1}\sqrt{e^{\sigma_2^2} - 1}}.$$

Observe that in general two log-normal random variables could not be made perfectly correlated, as the upper bound $\overline{\rho}$ cannot reach 1. If $\sigma_1 \to \infty$, the right bound converges to zero, i.e. $\lim_{\sigma_1 \to \infty} \overline{\rho}[X_1, X_2] = 0$. Similarly, for the left bound, $\lim_{\sigma_1 \to \infty} \underline{\rho}[X_1, X_2] = 0$ for $\sigma_1 \to \infty$.

A criticism of the Pearson correlation coefficient is that it only measures the linear relationship between two random variables, as illustrated in Example 2.24. Many alternative dependence measures are also used in the literature. For example, the Spearman correlation coefficient is a risk measure that can be used to describe the monotonic relationship, i.e. whether or not the realizations of two random variables tend to move in the same direction. As it is sufficient to use only the Pearson correlation coefficient in this book, interested readers are referred to Denuit et al. (2006) for details of Spearman and other correlation coefficients.

Appendix 2.A Proof of Equivalent Inequalities (2.1)

We shall provide a proof by contrapositive. The contrapositive of the statement $A \Rightarrow B$ is the opposite of $B \Rightarrow$ the opposite of A. If one can prove the contrapositive, then the original statement must also be true.

Let us first consider the "$\Leftarrow$". If $p \leq F(x)$, then $x \in A := \{x : F(x) \geq p\}$. It is clear from the definition of F^{-1} that $F^{-1}(p) = \inf A \leq x$.

For the "$\Rightarrow$" part, we consider the contrapositive of this equivalence. We want to show that $p > F(x) \Rightarrow F^{-1}(p) > x$. If $p > F(x)$, then by the right-continuity of F we can find a number $\epsilon > 0$ such that $p > F(x + \epsilon)$. Then by the supremum version of the definition of quantile function, we see that $F^{-1}(p) \geq x + \epsilon > x$. Now we have established the equivalence.

Appendix 2.B VaR as a Solution to a Minimization Problem

We define the function g and observe that

$$\begin{aligned} g(c) &:= \mathbb{E}[(X-c)_+] + c\epsilon = \mathbb{E}\left[\int_c^X 1dx\ I(X > c)\right] + c\epsilon = \mathbb{E}\left[\int_c^\infty I(X > x)dx\right] + c\epsilon \\ &= \int_c^\infty \mathbb{P}(X > x)dx + c\epsilon = \int_c^\infty \overline{F}(x)dx + c\epsilon, \end{aligned} \tag{2.17}$$

where $\overline{F}$ is the survival function of X. Setting the derivative of g with respect to c to zero yields

$$g'(c) = \epsilon - \overline{F}(c) = 0.$$

Since $g'(c)$ is an increasing function of c, it is clear that $c^* = \text{VaR}_{1-\epsilon}[X]$ is the global minimum of the function $g(c)$. In other words, $\text{VaR}_{1-\epsilon}$ is the minimizer of the optimal capital requirement problem (2.3).

Appendix 2.C Jensen's Inequality

Theorem 2.7 (Jensen's inequality) *If $v''(x) > 0$ (i.e. v is a convex function) and Y is a random variable, then*

$$\mathbb{E}[v(Y)] \geq v(\mathbb{E}[Y]).$$

Proof Let $\mu = \mathbb{E}[Y]$ and $D = \mathbb{E}[v(Y)] - v(\mu)$. We can rewrite D as

$$
\begin{aligned}
D &= \mathbb{E}\left[\int_{-\infty}^{Y} v'(y)\,\mathrm{d}y - \int_{-\infty}^{\mu} v'(y)\,\mathrm{d}y\right] \\
&= \mathbb{E}\left[\int_{\mu}^{Y} v'(y)\,\mathrm{d}y\right] \\
&\geq \mathbb{E}\left[(Y-\mu)v'(\mu)\right] \\
&= (\mu-\mu)v'(\mu) = 0.
\end{aligned}
$$

Problems

1. Suppose that two random variables X and Y are both normally distributed. Show that for any $p \in [1/2, 1]$,

$$
\mathrm{VaR}_p[X+Y] \leq \mathrm{VaR}_p[X] + \mathrm{VaR}_p[Y]. \tag{2.18}
$$

2. Let us consider pure endowment insurance, which is a life-contingent contract that pays a survival benefit upon the survival of the policyholder to maturity. Suppose that the survival benefit is \$100, the discount factor over the entire period to maturity is 0.9 and the probability of survival is 0.1. Then it is clear by the expected value premium principle that the pure premium should be given by $\$100 \times 0.9 \times 0.1 = 9$. Then the net liability from a single contract is given for any arbitrary policyholder i by

$$
L_i = \begin{cases} -9, & \text{with probability } 0.9; \\ 81, & \text{with probability } 0.1. \end{cases}
$$

The first term shows the profit from the premium \$9 and the second term is the present value $\$100 \times 0.9 - 9 = \$81$. Show that for some p,

$$
\mathrm{VaR}_p\left[\sum_{i=1}^{100} L_i\right] \geq \sum_{i=1}^{100} \mathrm{VaR}_p\left[L_i\right].
$$

In other words, this example shows that the VaR risk measure is in general not subadditive.

3. For a normal random variable X with mean μ and variance σ^2, show that

$$
\mathbb{E}[X|X>a] = \mu + \sigma\frac{\phi(\frac{a-\mu}{\sigma})}{1-\Phi(\frac{a-\mu}{\sigma})},
$$

and

$$\mathbb{E}[(X-a)_+] = \mu + \sigma \frac{\phi(\frac{a-\mu}{\sigma})}{1-\Phi(\frac{a-\mu}{\sigma})} - a\left(1 - \Phi\left(\frac{a-\mu}{\sigma}\right)\right).$$

4. The following expressions for the derivatives of risk measures are used in the derivation of capital allocation principles in Chap. 4.

   a. Show that for any continuous random variables X and S with joint distribution $f(s, x)$,

   $$\frac{\partial}{\partial h}\mathrm{VaR}_p[S+hX] = \mathbb{E}[X|S+hX = \mathrm{VaR}_p[S+hX]].$$

   Hint: Consider the expression

   $$\mathbb{P}(X+\alpha Y > \mathrm{VaR}[X+\alpha Y; p]) = 1-p,$$

   which can be rewritten as

   $$\int_{-\infty}^{\infty}\left[\int_{\mathrm{VaR}[X+\alpha Y;p]-\alpha y}^{\infty} f(x, y)\,\mathrm{d}x\right]\mathrm{d}y = 1-p.$$

   b. Show that for any continuous random variables X and S,

   $$\frac{\partial}{\partial h}\mathrm{TVaR}_p[S+hX] = \mathbb{E}[X|S+hX \geq \mathrm{VaR}_p[S+hX]].$$

5. Entropic risk measure is defined through the exponential utility function, i.e.

   $$\rho[X] = \frac{1}{\theta}\log\mathbb{E}[e^{-\theta X}].$$

   Suppose that X follows a normal distribution with mean μ and variance σ^2. Show the following results:

   a. The risk measure is given by $\rho[X] = \mu + \frac{1}{2}\sigma^2\theta$.
   b. The entropic risk measure is not coherent.

6. Given any two random variables X and Y, show the following results.

   a. The covariance of (X, Y) can be represented as

   $$\mathbb{C}(X, Y) = \int_{-\infty}^{\infty}\int_{-\infty}^{\infty}(\mathbb{P}(X \leq x, Y \leq y) - F_X(x)F_Y(y))\,\mathrm{d}x\,\mathrm{d}y.$$

   b. If X and Y are comonotonic, then $\mathbb{C}(X, Y) \geq 0$.

7. Consider X, Y to be two arbitrary random variables.

   a. Show that the regression coefficients a_R and b_R that minimize the squared distance

   $$\mathbb{E}[(Y - (aX + b))^2]$$

   are given by

   $$a^* = \frac{\mathbb{C}[X, Y]}{\sigma^2(X)}$$
   $$b^* = \mathbb{E}[Y] - a^*\mathbb{E}[X].$$

   b. Prove that

   $$\rho(X, Y)^2 = \frac{\sigma^2(Y) - \min_{a,b} \mathbb{E}[(Y - (aX + b))^2]}{\sigma^2(Y)}.$$

   c. Show that if X and Y are perfectly correlated, then (X, Y) must be comonotonic.

Chapter 3
Economics of Risk and Insurance

In this chapter, we provide an overview of four risk management techniques that are commonly used in practice, namely risk avoidance, risk retention, risk transfer, and risk mitigation. We then illustrate how these four techniques have been used in the business of insurance. The utility theory is introduced and used to explain why certain insurance product designs have been popular to meet the needs of risk management. As insurance is a scheme of risk sharing between two parties—an insurer and a policyholder—one can develop an act of balance in the framework of Pareto efficiency. Such a framework shall be used in later chapters to introduce other risk sharing schemes among multiple parties beyond traditional insurance.

3.1 Basics of Risk Management

As risks are unavoidable in all walks of life, everyone has to deal with risks in some way, consciously or unconsciously. In the following, we offer some examples of risk management techniques.

3.1.1 Risk Avoidance

Risk avoidance is an approach to manage a business by not engaging in certain risky activities. When a person is afraid of heights, he can simply choose not to climb mountains. A company may choose not to use certain hazardous materials to avoid any danger of improperly handling them. An entity with limited risk tolerance may adopt risk management policies to determine which risks to take on and which to avoid.

R. Feng, *Decentralized Insurance*, Springer Actuarial,
https://doi.org/10.1007/978-3-031-29559-1_3

3.1.2 *Risk Transfer*

Risk transfer is an approach that an entity takes to shift some risk to another party. A simple example of risk transfer is the responsibilities of a landlord and a tenant. While the property belongs to the landlord, the occupant has certain responsibilities to avoid damages to the property. If losses occur from damages due to negligence, then the tenant should be financially liable for repair costs.

Most risk transfer strategies are implemented through contractual obligations between two parties. A mathematical approach to understanding risk transfer strategies is to quantify and measure the amount of risk transferred from one party to another. In this section, we introduce the two most common types of risk transfer strategies.

Proportional Risk Transfer

The simplest approach is to transfer a stated percentage share of any potential loss from one party to another. In other words, the risk X under consideration can be split into two components

$$X = \alpha X + (1 - \alpha)X,$$

where the first component is considered as the ceded risk and the second as the retained risk. The proportional risk transfer is common in everyday life. For example, health insurance often requires coinsurance, under which the policyholder seeking health care pays for a portion of the healthcare service fee. When someone goes to see a doctor, the person and the insurer split the cost by a proportion. A research study (Manning et al., 1987), known as the RAND health insurance experiment, was done from 1974 to 1982 in the US to show that cost sharing reduces "inappropriate or unnecessary" healthcare costs. When patients were required to pay for 25% of their care (up to an out-of-pocket maximum that was based on their income), they reduced spending by about 17%. Similarly, when patients were required to pay for 95% of their care with the same controlled variables, the costs were smaller by about 30%. Participants with cost sharing made fewer medical visits and were admitted to hospitals less frequently. Today, coinsurance has become a common practice for cost control in healthcare insurance around the world. Note that in practice the cost of risk transfer should also be included. When taking the cost into account, we can rewrite the split as follows:

$$X = [\alpha X - p] + [(1 - \alpha)X + p],$$

where p is the cost of risk transfer. The expression inside the first set of parentheses is the net loss for the party receiving the risk transfer and that inside the second set of parentheses is the cost of the party ceding the risk.

Non-proportional Risk Transfer

Individuals and entities all have limited risk-bearing capacities, beyond which the consequences of potential losses can be catastrophic. A common approach for managing risk beyond one's risk-bearing capacity is to transfer the excessive risk to a third party. There are at least two ways of transferring excessive risks.

- *Ordinary deductible*
  For example, if the original risk is represented by the loss random variable X and the individual under consideration has a risk-bearing capacity of d, then the loss amount up to d is known as the retained risk, $X \wedge d = \min\{X, d\}$. For property and casualty insurance, the threshold d is also known as the (ordinary) deductible. The individual may choose to cede the excessive risk, $(X - d)_+ = \max\{0, X - d\}$, to a third party. The breakdown of the risk is hence given by

$$X = (X \wedge d) + (X - d)_+,$$

  where the first component is the retained risk and the second the ceded risk. It is often the case that the third party may not be able to or willing to take all of the excessive risks and can only take the risk up to its own risk-bearing capacity, say, u. Then the actual amount of coverage is given by $(X - d)_+ \wedge u$. In such a case, the ceding party is left with the risk $(X \wedge d) + [X - (u + d)]_+$. The breakdown of the original risk is then given by

$$X = (X \wedge d) + [X - (u + d)]_+ + (X - d)_+ \wedge u,$$

  where the first term and the second term represent the retained risk and the last term is the ceded risk to the third party. When such a risk transfer takes place from an insurer to a reinsurer, such an arrangement is often referred to as an excess-of-loss treaty.
- *Franchise deductible*
  A franchise deductible refers to the maximum amount of loss the ceding party has to pay before an insurer will pay in full. In contrast to the case of an ordinary deductible, the risk transferred under the franchise deductible is not just the excessive amount but rather the entire amount once the deductible is reached. If we use d to denote the franchise deductible, then the original risk is split into two components, where the individual pays for the loss when it is below the deductible and the insurer pays for the loss when it is above the deductible, i.e. $X = X_1 + X_2$ where

$$X_1 = \begin{cases} X, & \text{if } X \leq d; \\ 0, & \text{otherwise.} \end{cases} \qquad X_2 = \begin{cases} 0, & \text{if } X \leq d; \\ X, & \text{otherwise.} \end{cases}$$

  In other words, depending on the severity of loss, one party takes all.

3.1.3 Risk Retention

As an old saying goes, there is no such thing as a free lunch. All rewards come with certain risks. Every risk taker has to make a decision on whether or not to fully accept the risk. Sometimes, people undertake risks without thinking much about them. For example, when you walk on an uncovered street, the chance of being struck by lightning in a given year is around 1 in 500,000. A risk-averse person may choose to stay in during a rainy day, while an adventurous person goes out regardless. A decision, conscious or unconscious, is based on one's risk-bearing capacity. When the potential consequence is within the risk-bearing capacity, one can choose to accept the risk.

Risk retention is an individual's or organization's decision to take responsibility for a particular risk it faces. A typical approach to deal with risk retention is to "hope for the best and prepare for the worst". One can set up a "rainy day" fund which can be used later to absorb potential losses in worst scenarios in order to alleviate dire financial consequences.

In the business world, the same concept of a "rainy day" fund is formalized as a reserve. A risk retention strategy is reflected in a company's accounting books. In accounting, a balance sheet offers a snapshot of a company's financial condition. Take an insurer as an example. A balance sheet typically has two sides with an asset on one side and liability and net worth on the other side. The two sides are visualized in Fig. 3.1. Asset is a collection of items that an insurer owns. Liability

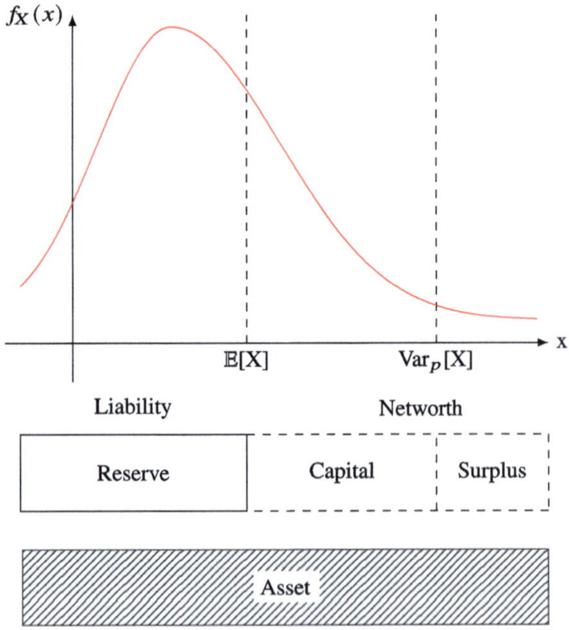

Fig. 3.1 Illustrative graph for the asset liability structure of an insurer

and net worth are a collection of items that an insurer owes to others. In simple terms, an insurer's financing side can be further broken down into reserve, capital, and surplus. Reserve measures the amount that an insurer sets aside to meet future insurance liability, i.e. benefit obligations to policyholders. The amount of reserve is in principle determined by the insurer's expected losses. Capital refers to the additional amount that an insurer keeps in order to absorb unexpected losses. While there are many ways of determining reserves and capitals, one approach is to set a risk capital as a quantile of the underlying risk. In other words, capital is set to ensure that sufficient funding is available to pay for losses with a certain level of confidence. In general, the more risk is retained by an insurer, the higher the reserve and capital to be set aside on an insurer's balance sheet. As the two sides have to balance, the excessive amount of assets held by an insurer beyond reserve and capital is often referred to as surplus, representing the insurer's profits and shareholders' equity. In summary, these accounting items are connected through these identities:

$$\begin{aligned}\text{Liability} + \text{Net Worth} &= \text{Reserve} + \text{Capital} + \text{Surplus} \\ &= \text{Expected Losses} + \text{Unexpected Losses} + \text{Profits}.\end{aligned}$$

We shall explore the calculations of reserve and capital further in Chap. 4.

3.1.4 Risk Mitigation

Risk mitigation refers to the collection of strategies that can be used to lessen the effects of negative consequences from potential losses. For example, to reduce the potential fire damage in a house, it is a common practice to install smoke detectors, stock fire extinguishers in multiple places, etc. Such risk mitigation strategies change the loss distribution of the underlying risk by reducing the frequency or severity of losses. However, in this book, we are only concerned with financial arrangements such as insurance. Therefore, we only consider a class of risk mitigation techniques known as risk diversification.

The fundamental principle underlying risk diversification is the so-called law of large numbers. In a homogeneous risk group, where all losses are independent and identically distributed (i.i.d.), the average loss converges to the theoretical mean of each loss random variable.

Theorem 3.1 (Strong law of large numbers) *Let $X_1, X_2, \cdots, X_n$ be an infinite sequence of i.i.d. random variables with finite mean $E(X_1) = E(X_2) = \cdots = \mu$. Then*

$$\overline{X}_n := \frac{1}{n}(X_1 + \cdots + X_n) \to \mu, \qquad n \to \infty.$$

In probability theory, the convergence $\rightarrow$ is known as an almost sure convergence. We offer here a non-technical interpretation. Observe that $(\overline{X}_1, \overline{X}_2, \cdots, \overline{X}_n, \cdots)$ is an infinite series of random variables. Every realization of this sequence of random variables is a series of real numbers, $(x_1, x_2, \cdots, x_n, \cdots)$. The probability that such a series converges to μ is one. The significance of this principle is that, while individual losses are hard to predict, the average loss is predictable and manageable. Some losses could be severe, and some may be negligible. In a large risk group, the losses can balance out and the average is close to its theoretical mean. The certainty increases with the size of the risk group. In the following, we can visualize the result of the law of large numbers.

Example 3.1 Suppose we intend to test the fairness of a coin, i.e. whether the coin has an equal chance of landing on a head or a tail. We can do so by counting the number of heads in a sequence of coin tosses. The number of heads in each toss is a Bernoulli random variable, denoted by X_1. Let p be the probability of a head and q be the probability of a tail. Note that these parameters are unknown before our experiments. Then its probability mass function is given by

$$\mathbb{P}(X_1 = x) = \begin{cases} p, & x = 1, \\ q, & x = 0. \end{cases}$$

Let X_k be the number of heads in the k-th coin toss, $k = 1, 2, \cdots, n$. Count the total number of heads after n coin tosses:

$$S_n := \sum_{k=1}^{n} X_k.$$

Then it is easy to show that S_n is a binomial random variable with parameters n and p, and its probability mass function is given by

$$\mathbb{P}(S_n = x) = \binom{n}{x} p^x q^{n-x}, \qquad x = 0, 1, \cdots, n.$$

Consider the sample average

$$\overline{X}_n := \frac{1}{n} S_n.$$

According to the law of large numbers, we can expect that the sample mean converges to the chance of landing on a head, i.e.

$$\overline{X}_n \rightarrow p.$$

As we toss the coin more and more, the percentage of heads in all counts is expected to be closer and closer to the probability of a head. If this percentage is close to $1/2$, then we can believe that this is a fair coin. As shown in Fig. 3.2, the

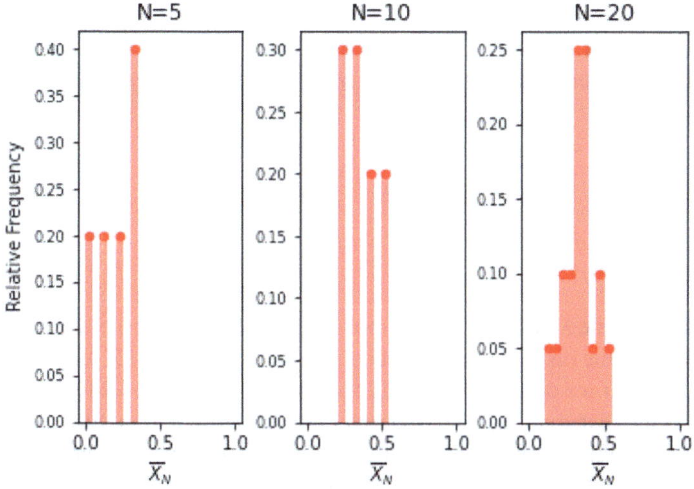

(a) Small n

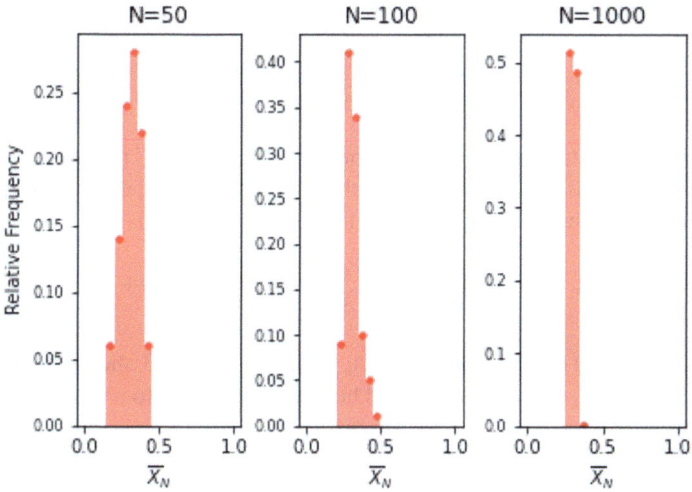

(b) Large n

Fig. 3.2 Percentage of heads in coin tosses

percentage of heads is close to 0.3, which indicates that the coin is more likely to land on a tail than on a head.

Another fundamental principle often used for risk diversification is the central limit theorem, which states that the sample mean is asymptotically normally distributed.

Theorem 3.2 (Lindeberg-Lévy central limit theorem) *Let $X_1, X_2, \cdots$, be an infinite sequence of i.i.d. random variables with finite mean and variance. Then*

$$Z_n := \frac{\overline{X}_n - \mathbb{E}(\overline{X}_n)}{\sqrt{\mathbb{V}(\overline{X}_n)}} \to Z, \qquad n \to \infty,$$

where Z is a standard normal random variable.

In the above-stated theorem, the symbol $\to$ indicates a convergence in distribution. In simple words, the distribution of Z_n becomes closer to that of Z as n increases. It is often the case that we do not necessarily have a parametric model for the underlying risk. Even if the distributions of individual risks are known, it is difficult to compute the distribution of the aggregate risk. The central limit theorem can be used to provide an approximate distribution of the aggregate risk.

Example 3.2 Let us continue with the coin experiment in Example 3.1. The Central Limit Theorem tells us that the estimator $\overline{X}_n$ is asymptotically normal, i.e.

$$Y_n := \sqrt{n}\frac{\overline{X}_n - p}{\sqrt{pq}} \to \mathcal{N}(0, 1),$$

where $\mathcal{N}(0, 1)$ is a standard normal random variable and "$\to$" refers to convergence in distribution. Let us verify numerically the conclusion of the central limit theorem. One can show that the exact probability mass function of Y_n is given by

$$\mathbb{P}(Y_n = y) = \binom{n}{h} p^h q^{n-h}, \qquad h := \sqrt{npq}\,y + np,$$

where $y = (k - np)/\sqrt{npq}$ and $k = 0, 1, \cdots, n$. Figure 3.3 shows the probability mass functions of Y_n for various n's and how they converge to a normal distribution as n increases.

As fundamental principles in the insurance business, applications of the law of large numbers and the central limit theorem have been discussed in a large volume of papers. To name just a few, readers can find examples in Cummins (1991) and Powers and Shubik (1998), and for equity-linked insurance and annuities in Feng and Shimizu (2016).

3.2 Basics of Insurance

In this section, we review the building blocks of traditional insurance and where insurance has been used as an important economic mechanism for risk management.

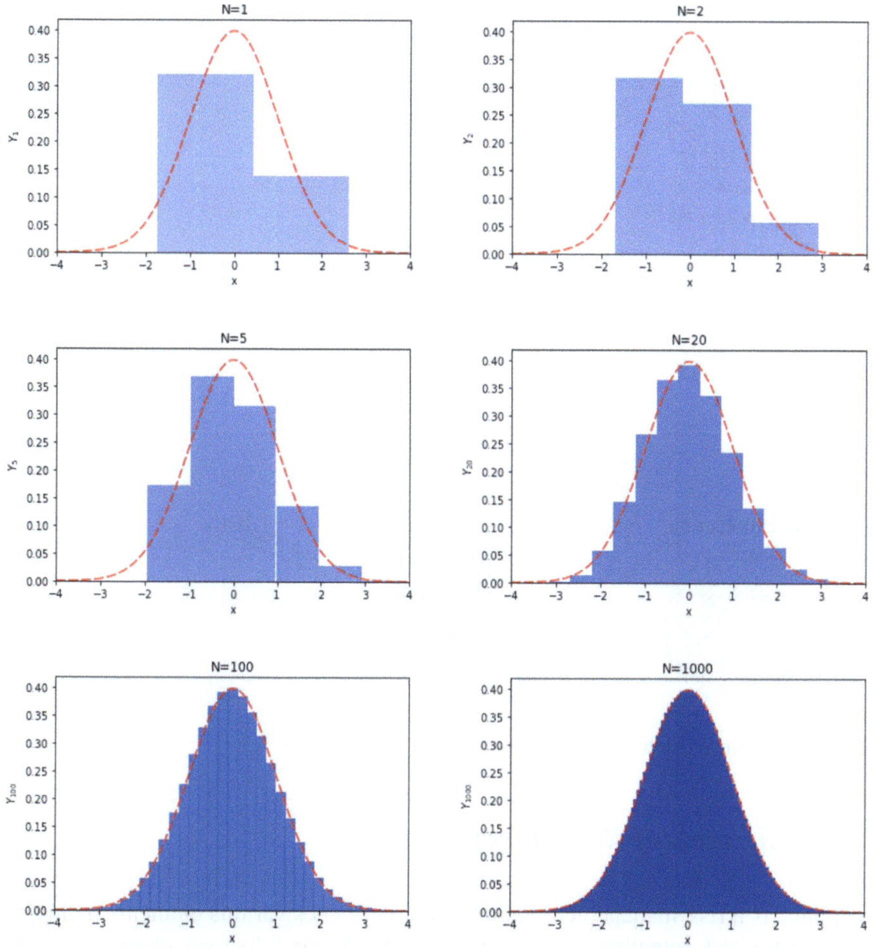

Fig. 3.3 Convergence of the probability function of sample mean to normal density

An insurance policy is typically a bilateral contract between an insurer and a policyholder. The insurer agrees to provide compensation to the policyholder in a loss event, in exchange for a fixed premium from the policyholder. In other words, the policyholder buys some certainty and passes on some uncertainty to the insurer. While each insurance policy is bilateral, the identical contract is sold to tens of thousands of policyholders. The insurer is in essence the risk taker from a large pool of individual risks from policyholders. The insurer uses a range of risk management tools to assess, control, and manage risks and profit from the business of dealing with risks.

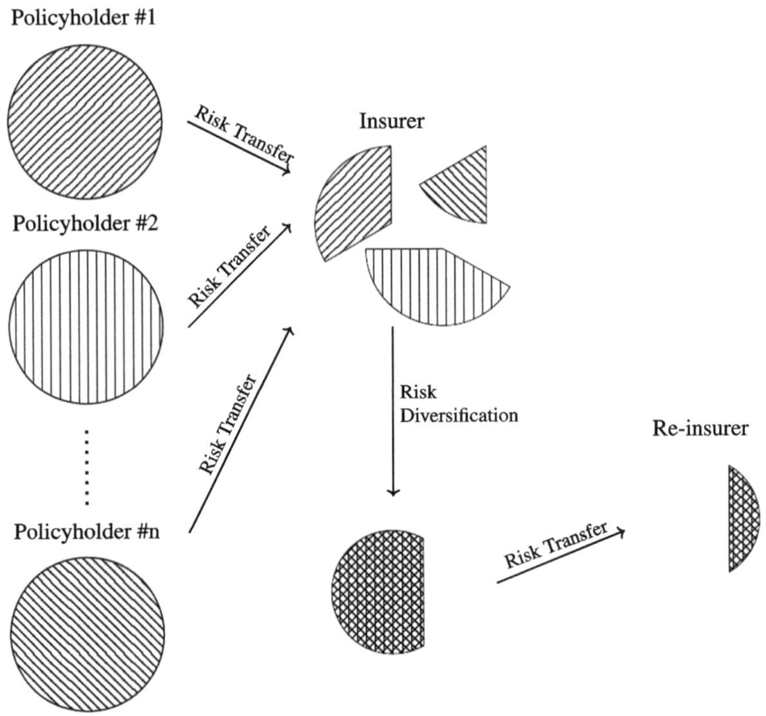

Fig. 3.4 Illustrative graph for risk management tools of an insurer

3.2.1 *Elements of Risk Management*

Insurance is a risk management solution that combines various elements including risk transfer, risk retention, and risk mitigation. Figure 3.4 shows elements of risk management in a typical insurance business.

Risk Transfer

Insurance is often a bilateral contract, known as a policy, between a policyholder and an insurer. The same insurance policy is sold to tens of thousands of policyholders. Under each policy, a policyholder pays a fixed premium in advance and in exchange the insurer agrees to provide compensation for each policyholder's loss. In other words, each policyholder transfers some risk to the insurer.

Risk Mitigation

The quintessence of an insurance business is the pooling of funds from a large number of policyholders to pay for losses incurred by a few. Because an insurer can pool together risks from tens of thousands of policyholders, it can develop an economy of scale by the law of large numbers and make the cost of coverage predictable and manageable.

Example 3.3 (*Mortality Risk—Term Life Insurance*) Consider a pure endowment insurance, which pays \$100 should the policyholder survives 10 years. Assume that

- the 10-year discount factor is 0.9;
- the probability of survival is 0.1.

We can determine the net premium by $100 \times 0.9 \times 0.1 = 9$. Observe that the present value of loss from a single contract is given by

$$L_i = \begin{cases} 81, & \text{with prob } 0.1; \\ -9, & \text{with prob } 0.9. \end{cases}$$

For the standalone contract, there is a 10% chance of incurring a loss nine times the premium.

Now, suppose that the same contract is sold to 1000 policyholders. The average loss for each contract is

$$\overline{L} = \frac{1}{1000} \sum_{i=1}^{1000} L_i,$$

which has mean 0 and variance 0.729. The variation is so small that the probability of having a loss the size of the premium is zero.

Risk Retention

For any potential loss beyond its own risk-bearing capacity, the insurer can pass on the excessive amount to a reinsurer or engage in some hedging strategy. The insurer typically sets up reserve and capital to absorb potential losses from the retained risk portfolio.

Example 3.4 (*Insurance*) An insurance company has n clients. The kth client has insurance claim X_k. Suppose that all insurance claims are independent and identically distributed with the same mean μ and variance σ^2. The total loss of an insurer is given by

$$S_n = \sum_{k=1}^{n} X_k.$$

How much money does the insurer have to hold in order to have sufficient funding to cover losses with a certain probability? By the Central Limit Theorem, we know that

$$\frac{S_n - \mu n}{\sigma \sqrt{n}}$$

is approximately a standard normal random variable. To ensure that it has sufficient capital to cover its liability, the insurer can set aside the amount K such that

$$\mathbb{P}(S_n \leq K) \geq \alpha.$$

Therefore, the amount K is given by

$$K = \mu n + \sigma\sqrt{n}\Phi^{-1}(\alpha).$$

Observe that the amount per contract decreases with the number of contracts, i.e.

$$\mu + \frac{\sigma}{\sqrt{n}}\Phi^{-1}(\alpha)$$

reserve + capital.

In other words, the more policyholders there are, the lower the cost of coverage. When n is sufficiently large, the amount to set aside is about the same as the expected loss. This result is consistent with the discussion in Sect. 3.1.3. The amount to set aside can be viewed as the sum of two components—reserve and capital, where the reserve reflects the expected loss and the capital protects the insurer from insolvency in rare events of extreme loss.

3.3 Economics of Risk and Insurance

In this section, we present the economic theory of risk and insurance and explain how various forms of risk sharing have been justified as optimal risk management solutions. Let us begin with a thought experiment.

Deal or No Deal is a popular American television game show (see Fig. 3.5). In each game, a contestant is presented with 26 briefcases, each of which contains a value amount between \$0.01 and \$1, 000, 000. The contestant is asked to set aside one of the 26 briefcases at the beginning. Over the course of the game, the contestant

Fig. 3.5 American TV show *Deal or No Deal*. *Source* Wikimedia Commons

is asked to eliminate a briefcase at a time and the cash amount in each eliminated briefcase is revealed to the audience. From time to time, the contestant receives a call from a back-stage banker who offers the contestant a deal to take some cash and leave the game. The contestant has to choose between a fixed amount of cash and continuing with the game and facing the uncertainty of outcomes. When a contestant refuses all deals and is left with only one briefcase, he or she is given another chance to trade the one left on the stage for the one set aside at the beginning. The contestant is given the cash in the chosen briefcase at the end. The show is widely popular as it is all about personal choices facing uncertainty and dramatic financial consequences.

We can play a simplified version of the Deal or No Deal game. Suppose that you are presented with ten briefcases containing the following cash amounts. However, the briefcases are closed and the cash amounts inside are not disclosed at the beginning.

Briefcase	?	?	?	?	?	?	?	?	?	?
Cash	\$0	\$0	\$0	\$10, 0000, 000	\$0	\$0	\$0	\$0	\$0	\$0

You can pick one out of the 10 briefcases and take home the cash amount in the chosen briefcase. For 9 out of 10 briefcases that you can choose, you will receive nothing. You can only bet on the one briefcase for which you can take home 10 million dollars. Just when you are ready to pick a number, you receive a call from a banker who gives you a chance to either take a cash offer of $\$1,000,000$ or to open the briefcase of your choice with an unknown amount. Would you take the banker's offer? In other words, are you better off taking the guaranteed one million dollars or trying your luck between nothing and ten million dollars?

If one can play the game for unlimited time, then the two choices make no difference. Let us consider the average of outcomes:

- No Deal: $(9/10) \times 0 + (1/10) \times 10,000,000 = 1,000,000$;
- Deal: $1,000,000$.

From the viewpoint of the host, this is a good estimate of the cost per game in a series of episodes. However, as a participant, you only have one shot. Despite the same outcome in terms of averages, the impact of the two decisions on your personal wealth can be drastically different. In fact, your decision reflects your risk attitude. If you choose no deal, you are a risk-seeking person. Otherwise, you are a risk-averse person. In the next section, we shall reveal why that is the case and how risk aversion plays a role in decision-making.

3.3.1 *Utility*

Utility theory is used in economics to explain how economic agents behave based on the premise that they can rank different choices by preference measures and take actions according to their preferences. We can rank an individual's preferences using a mathematical tool, $u(x)$, known as a *utility function*, which is a measure of satisfaction from a certain level of wealth x. It possesses two properties:

- (Non-decreasing) $u(x) \leq u(y)$ if $x \leq y$.
- (Diminishing marginal utility/risk aversion) $u''(x) < 0$.

The first property is obvious, as, in general, the more wealth an individual possesses, the better off is the person's financial condition. The second property shows that the marginal benefit of additional wealth decreases as an individual becomes richer. For example, offering $20 dollars to a beggar could be a day of food and drinks but the same amount does not affect much at all to the life of a billionaire who lives a lavish lifestyle. We often assume that a rational economic agent makes decisions to maximize his/her overall utility.

In the context of risk management, we often consider decisions with respect to potential losses, where it is sometimes more convenient to consider a *disutility function*. For example, one could define a disutility function as $v(x) = u(w - x)$ where w is the initial wealth of an individual. With the analogy to the utility function, a disutility function is non-increasing and has increasing marginal disutility.

Multi-attribute utility functions are also used in utility theory to describe an economic agent's preference over a bundle of attributes. For example, one's satisfaction with a job may depend on both the salary and the number of vacation days. The multi-attribute utility is often used to describe decision-making with regards to the allocation of different resources.

Example 3.5 (*Utility functions*) There are a number of utility functions that are commonly used in economic analysis:

- Linear utility: $u(x) = x$;
- Quadratic utility: $u(x) = -(\alpha - x)^2, x \leq \alpha$;
- Logarithmic utility: $u(x) = \ln(\alpha + x), x > -\alpha$;
- Exponential utility: $u(x) = -\alpha e^{-\alpha x}, \alpha > 0$;
- Power utility: $u(x) = x^c, x > 0, 0 \leq c \leq 1$.

One can easily verify that they all satisfy the two defining properties of the utility function.

Some common examples of multi-attribute utility functions include

- Linear utility: $u(x_1, x_2) = ax_1 + bx_2, a, b > 0$;
- Cobb-Douglas utility: $u(x_1, x_2) = Ax_1^a x_2^b$ for some $A > 0$.

When facing uncertainty, it is assumed that a rational economic agent makes decisions to maximize expected utility. Therefore, we shall compare the consequences of two actions which are measured by the agent's expected utility. Let us revisit the Deal or No Deal thought experiment. Take an exponential utility function as an example.

$$u(x) = 1 - e^{-x}.$$

Let X be the money you take away at the end of the game. Consider the expected utility from your two choices.

- No Deal (opening one of the ten cases):
  Under the best scenario where you happen to open the 10 million dollar briefcase, your utility is given by $u(10,000,000) \approx 1$. Under the worst scenario where you open one of the other nine briefcases, you receive nothing and your utility is given by $u(0) = 0$. Combining both cases with the chances, we can determine the expected utility by

$$\mathbb{E}[u(X)] = \frac{9}{10} \times 0 + \frac{1}{10} \times 1 = \frac{1}{10}.$$

- Deal (taking the banker's offer):
  There is no uncertainty with the banker's deal. Your utility is given by

$$u(1,000,000) \approx 1.$$

It is clear from this calculation that, when the utility is considered, an individual would choose to take the deal. When you have nothing to begin with, you are already very well off with having a certain cash reward of one million. The additional utility of earning from one million to ten million is not high enough for you to risk losing the utility from a guaranteed one million reward.

3.3.2 Risk Aversion

An economic agent that makes choices based on the maximization of expected utility of the type described earlier is considered to be risk-averse. The reason is that the agent would always choose the option with certainty than the one with uncertain outcomes when the two options have the same mean. The result is a direct consequence of Jensen's inequality in Appendix 2.C.

$$\mathbb{E}[u(w - X)] \leq u(\mathbb{E}[w - X]) = u(w - \mathbb{E}[X]).$$

This result shows the concept of risk aversion, meaning that a decision maker would always prefer a fixed outcome $\mathbb{E}[X]$ over the uncertain outcome X.

While we use utility to understand preference, it is also important to remember that the underlying utility function does not need to be uniquely determined. Observe that

$$\mathbb{E}[u(w - X)] \leq \mathbb{E}[u(w - Y)]$$

is equivalent to

$$\mathbb{E}[u^*(w - X)] \leq \mathbb{E}[u^*(w - Y)],$$

where $u^*(x) = au(x) + b$ is considered an affine transform of $u(x)$ for arbitrary positive constant a and real constant b. Therefore, we often use utility functions for which $u(0) = 0$ and $u'(0) = 1$ for simplicity.

In general, the higher curvature of the utility function, the stronger the risk aversion. However, we cannot use u'' alone to measure the tendency of risk aversion, as any affine transform of a utility function implies the same preference order as the original utility function. Therefore, we often use a measure that stays constant with an affine transform, which is known as the coefficient of absolute risk aversion

$$\gamma(x) = \frac{-u''(x)}{u'(x)}.$$

A particular case of interest is the exponential utility function, which has a constant absolute risk aversion $\gamma(x) = \alpha > 0$. Hence, the parameter α is referred to as the risk aversion rate for exponential utility.

3.3.3 Optimal Risk Transfer—Non-proportional Insurance

In a way similar to the earlier thought experiment, we can use the utility theory to compare the consequences of economic agents' actions for risk transfer. We want to explain why some common forms of risk transfer are preferred in some circumstances.

Suppose an individual has a non-negative risk X and is willing to pay a fixed premium P in exchange for ceding a portion $I(X)$ to an insurer. Assume that $0 \leq I(x) \leq x$ for all $x \geq 0$.

$$X = I(X) + R(X),$$

where her retained risk is denoted by $R(X)$. Suppose that the premium P is given by $P = \mathbb{E}[I(X)]$. The natural question to ask is what is the optimal form of risk transfer that minimizes the individual's variance, i.e. $\mathbb{V}(R(X))$. The answer is known to be an excess-of-loss insurance policy

$$I^*(x) = (x - d)_+,$$

where the deductible is determined by

$$\mathbb{E}[(X - d)_+] = P.$$

The optimal risk transfer is visualized in Fig. 3.6. The reasoning is that, in order to achieve the smallest variance, the individual should retain small losses and pass on large losses to the insurer. In other words, one should set up a threshold d such that all losses below d are retained and any excessive amount should be carried by an insurer.

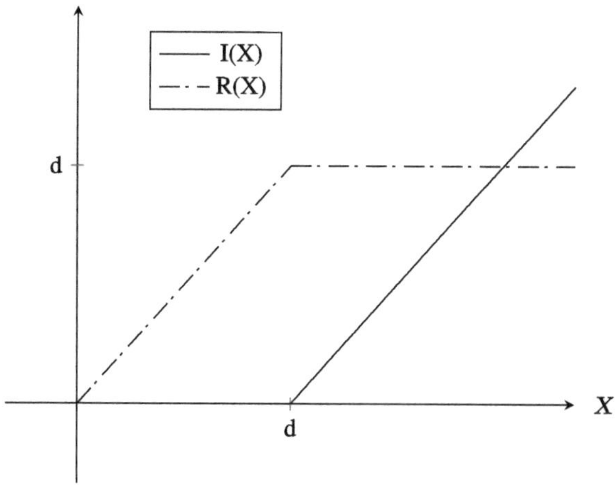

Fig. 3.6 Excess-of-loss insurance

To prove this, consider an arbitrary risk transfer plan for which the retained risk is denoted by $R(X)$. Note that the optimal retained risk is $R^*(X) = X \wedge d$. Because $\mathbb{E}[R^*(X)] = \mathbb{E}[R(X)]$, we only need to show that $\mathbb{E}[R^*(X)^2] \leq \mathbb{E}[R(X)^2)]$, which is equivalent to show that $\mathbb{E}[(R^*(X) - d)^2] \leq \mathbb{E}[(R(X) - d)^2]$. When $x \geq d$, observe that

$$|R(x) - d| \geq 0 = |R^*(x) - d|.$$

When $x < d$, it is clear that $R(x) < d$ since $R(x) \leq x$. Therefore,

$$|R(x) - d| = -(R(x) - d) \geq -(x - d) = |R^*(x) - d|.$$

Therefore, $\mathbb{E}[(R(X) - d)^2] \geq \mathbb{E}[(R^*(X) - d)^2]$. The proof is complete.

In fact, the excess-of-loss insurance not only minimizes the insured's variance but also maximizes her expected utility. In other words,

$$\mathbb{E}[u(w - R(X)] \leq \mathbb{E}[u(w - R^*(X))].$$

Recall that $u''(x) < 0$. Thus,

$$u(w) - u(x) = \int_x^w u'(v)\, \mathrm{d}v \leq (w - x)u'(x).$$

Note that $R(X) \leq d$ and u' is a non-increasing function:

$$\mathbb{E}[u(w - R(X)] - \mathbb{E}[u(w - R^*(X))] \leq \mathbb{E}[(R^*(X) - R(X))u'(w - R^*(X))]$$
$$\leq \mathbb{E}[(R^*(X) - R(X))]u'(w - d) = 0.$$

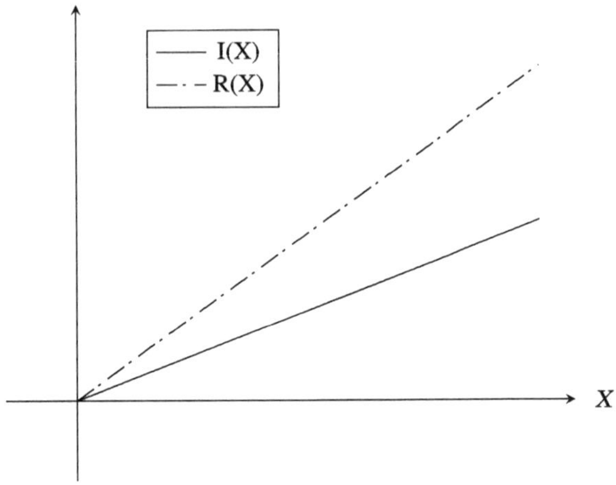

Fig. 3.7 Proportional insurance

Therefore, a utility-maximizing economic agent would always prefer the excess-of-loss policy as the optimal form of risk transfer given a fixed premium.

3.3.4 Optimal Risk Transfer—Proportional Insurance

Let us consider another scenario in which proportional insurance arises as the optimal insurance. Consider the risk exchange between an insurer and a reinsurer. The insurer acquires an insurance business with the risk X. As an insurer also has limited risk-bearing capacity, the risk from its own product portfolio can be split into

$$X = I(X) + R(X),$$

of which the insurer is willing to retain a certain portion of the risk $R(X)$ up to a fixed variance, i.e.

$$\mathbb{V}[R(X)] = V.$$

The rest $I(X)$ is ceded to a reinsurer. What type of reinsurance contract should be offered in order for the reinsurer to minimize its variance?

The answer is proportional insurance. Figure 3.7 shows a visualization of the optimal risk transfer as a proportional insurance. Observe that

$$\mathbb{V}[I(X)] = \mathbb{V}[X] + \mathbb{V}[R(X)] - 2\mathbb{C}[X, R(X)].$$

Then it is clear that the variance is minimized if X and $R(X)$ are linearly dependent, i.e. $R(X) = \beta + \alpha X, \alpha > 0$. Since $0 \leq R(x) \leq x$, we must have $\beta = 0$. Hence, $\mathbb{V}[R(X)] = \alpha^2 \mathbb{V}[X] = V$, which in turns implies that

$$\alpha = \sqrt{\frac{V}{\mathbb{V}[X]}}.$$

In summary, the optimal risk transfer between an insurer with a tolerance of uncertainty up to a fixed variance and a reinsurer that seeks to minimize its own variance is given by

$$R(X) = \alpha X, \qquad I(X) = (1 - \alpha)X.$$

As we shall show in later chapters, both proportional and non-proportional risk transfers are used frequently in both traditional and decentralized insurance.

3.4 Pareto Optimality

As alluded to in the earlier discussions, we often face the issue of an allocation of risks among different participants, whether between a policyholder and an insurer, or between an insurer and a reinsurer. A common approach to deal with the allocation of resources or uncertainty is known to be Pareto optimality.

Definition 3.1 (*Pareto Optimal*) A situation is Pareto optimal if it is not possible to make one person better off without another worse off.

The concept is named after Vilfredo Pareto, an Italian economist, who used the concept in his study of economic efficiency. The implicit assumption for Pareto optimality is that a social planner makes an allocation of limited resources based on participants' well-being. A Pareto optimal solution is an allocation strategy where no alternative can be available to improve one participant's well-being without sacrificing that of someone else.

Let us consider an intuitive example of Pareto optimality. An economy contains two participants and two goods, apples and bananas. Participant A likes apples and dislikes bananas. In other words, the person's well-being is improved with more apples or fewer bananas. On the contrary, participant B likes bananas and dislikes apples. There are a limited number of apples and bananas available in the economy. The only allocation that is Pareto optimal is that in which participant A has all the apples and participant B has all the bananas. For any other allocation, one person has some units of disliked goods, and would be better off if those units are given to the other person.

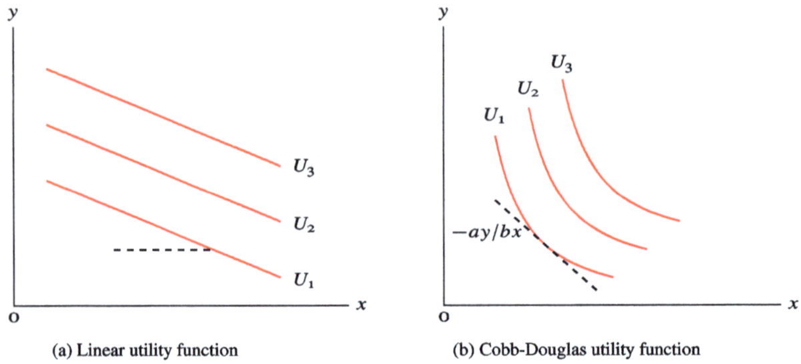

Fig. 3.8 Utility functions and their indifference curves

3.4.1 Marginal Rate of Substitution

We generalize this idea with a simple exchange economy. Assume that there are two consumers, A and B, and two types of goods. The total supplies of the two goods are fixed at s and t, respectively. The amounts of goods acquired by consumer A are denoted by x^A and y^A, and the amounts acquired by consumer B are denoted by x^B and y^B. We typically use the utility function as a measure of consumers' well-being. It is clear from the discussion that each consumer's well-being depends on both types of goods. Suppose that the utilities of the two consumers are given by $U_A(x^A, y^A)$ and $U_B(x^B, y^B)$, respectively. A common approach to understanding the consumers' preferences is to identify a set of indifference curves. An indifference curve (IC) is a set of bundles between which the consumer is indifferent. The farther the curve is away from the origin, the higher the consumer's utility is. For example, in Fig. 3.8a, the consumer under consideration is assumed to have a linear utility $U(x, y) = ax + by$. The ICs U_1, U_2, U_3 represent all bundles of (x, y) with the same utility, respectively. The utility of the consumer for all bundles in U_3 is higher than that for all bundles in U_2, and the utility for U_2 is higher than that for U_1. Similarly, we show the ICs for a consumer with the Cobb-Douglas utility function in Fig. 3.8b. The IC farthest away from the origin has the highest utility among the three ICs.

The slope of a tangent line to the IC is called the marginal rate of substitution (MRS), which measure the consumer's willingness to trade one good for another. The slope can be calculated as follows. The graph of an IC is given by $\{(x, y(x))|U(x, y(x)) = c\}$ for some constant c. Observe that

$$0 = \frac{\mathrm{d}c}{\mathrm{d}x} = \frac{\mathrm{d}U(x, y(x))}{\mathrm{d}x} = U_x(x, y(x)) + U_y(x, y(x))\frac{\mathrm{d}y(x)}{\mathrm{d}x},$$

where U_x and U_y are the partial derivatives of the utility function U with respect to x and y. Therefore, the slope of a tangent line to the IC is given by

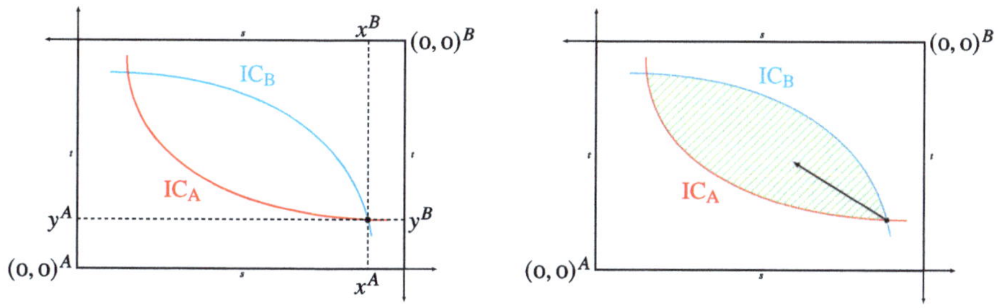

(a) Edgeworth Box (b) Edgeworth Box Improvement

Fig. 3.9 Edgeworth box

$$\frac{\mathrm{d}y(x)}{\mathrm{d}x} = -\frac{U_x(x, y(x))}{U_y(x, y(x))}.$$

If the consumer has the amount x of the first good and the amount y of the second good, then the MRS is defined as

$$\text{MRS} = -\frac{U_x(x, y)}{U_y(x, y)}.$$

For example, the MRS of the consumer with the linear utility function in Fig. 3.8a is given by $-a/b$ and the MRS of the consumer with the Cobb-Douglas utility function is $-ay/bx$, described in Fig. 3.8b.

3.4.2 *Pareto Efficiency*

An Edgeworth box is a graphical representation that helps to analyze the exchange economy of two goods between two people. It is a convenient way to depict the preferences of two consumers in one diagram. We first draw A and B's indifference curve in the same way as in Fig. 3.8, and turn B's upside down to overlay it on A's. For example, in Fig. 3.9a, a consumer A has x^A units of the first good and y^A units of the second good. A consumer B has x^B units of the first good and y^B units of the second good. The size of a box represents the total amounts of the goods in the economy, i.e. there are $x^A + x^B = s$ of the first good and $y^A + y^B = t$ of the second good in total.

An allocation is a distribution of the total amount of goods to all consumers. Any point in the Edgeworth box represents an allocation of the two goods between the two consumers. For example, the red curve in Fig. 3.9 represents a set of bundles of the two goods that gives the same utility to consumer A and the blue curve represents a set of bundles with the same utility for consumer B. In an exchange economy, an allocation is *Pareto-Efficient* if one cannot find another allocation that makes all consumers at

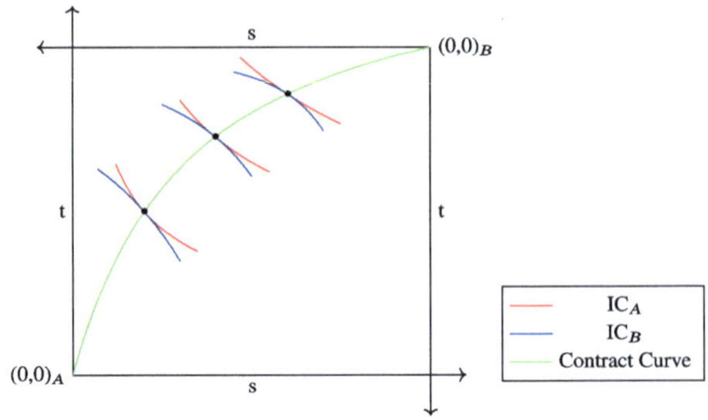

Fig. 3.10 Contract curve in an Edgeworth box

least as well off and makes at least one consumer strictly better off. Note that, in Fig. 3.9b, any point to the right of IC_A leads to a higher utility for consumer A and any point to the left of IC_B yields higher utility for consumer B. Therefore, any point in the shaded area represents an improvement for both consumers. So, reallocating the resources through the direction that the arrow points toward can make both better off. As one can imagine, as we move IC_A upward (different ICs with higher utility for A) and move IC_B downward (different ICs with higher utility for B), the shaded area shrinks and both consumers are better off with any point in the shaded area. We keep moving the two representative ICs until the two lines intersect each other at only one point, also known as a tangency point. Such a point is Pareto-efficient, because a departure would make at least the two consumers worse off. Note, however, the two ICs shown in Fig. 3.9a are arbitrary, and hence such Pareto-efficient points are not unique.

The contract curve is the set of all Pareto-efficient allocations in the exchange economy. With well-behaved indifference curves, the contract curve is the set of tangency points between A's indifference curves and B's indifference curves (see Fig. 3.10). Recall that the slope of tangent lines to ICs is given by the marginal rates of substitution. Therefore, the Pareto-efficient allocations can be determined by

$$\text{MRS}^A = \text{MRS}^B. \tag{3.1}$$

Example 3.6 Suppose that consumers A and B share the apples and bananas, where the total amount of apples and bananas are s and t, respectively. Assume that consumers A and B have Cobb-Douglas utility such that

$$U_A(x_A, y_A) = x_A y_A^2, \quad U_B(x_B, y_B) = x_B^2 y_B.$$

Note that the MRS of A and B are

$$\text{MRS}^A = -\frac{y_A}{2x_A}, \quad \text{MRS}^B = -\frac{2y_B}{x_B},$$

i.e. A is willing to trade $y_A/2x_A$ of apple for one banana, and B is willing to trade $2y_B/x_B$ of apple for one banana. When $\text{MRS}_A = \text{MRS}_B$, the Pareto optimal points are achieved. Since $x_A + x_B = s$ and $y_A + y_B = t$, we can obtain the solution

$$\frac{y_A}{2x_A} = \frac{2y_B}{x_B} \Longleftrightarrow \frac{y_A}{2x_A} = \frac{2(t - y_A)}{(s - x_A)} \Longleftrightarrow y_A = \frac{4x_A t}{3x_A + s}, \quad y_B = \frac{x_B t}{-3x_B + 4s}.$$

Keep in mind that in this case, Pareto optimal solutions are not unique since $0 \le x_A, x_B \le s$ are arbitrary. The contract curve for consumer A is given by $\{(x_A, y_A)|y_A = 4x_A t/(3x_A + s)\}$ and that for consumer B is given by $\{(x_B, y_B)|y_B = x_B t/(-3x_B + 4s)\}$.

3.4.3 *Risk Allocation*

In the context of insurance and risk management, we often need to consider the distribution of risks instead of assets. The concept of Pareto efficiency can be readily extended to risk management problems.

Example 3.7 Suppose that there are two economic agents A and B, both of whom have exponential utility functions

$$u_i(x) = \frac{1}{\gamma_i} e^{-\gamma_i x}, \qquad i = A, B,$$

where γ_i is the risk aversion rate of agent $i = A, B$. The two agents reach some profit/loss sharing agreement where each party carries a portion of the profit/loss from the other party. Let $\alpha \in (0, 1)$ be the self-retained portion of agent A's own risk and $\beta \in (0, 1)$ be the portion of agent B's risk carried by agent A. Thus, agent B carries a portion $1 - \alpha$ of agent A's risk and a portion $1 - \beta$ of its own risk. The profit/loss X_i of agent i is normally distributed with mean zero and variance $\sigma_i^2 > 0$. Therefore, the two agents' expected utilities are given by

$$\begin{aligned}
U^A(\alpha, \beta) &= \frac{1}{\gamma_A} \mathbb{E}\Big[\exp\{-\gamma_A(w_A - \alpha X_A - \beta X_B)\}\Big] \\
U^B(\alpha, \beta) &= \frac{1}{\gamma_A} \mathbb{E}\Big[\exp\{-\gamma_B(w_B - (1 - \alpha)X_A - (1 - \beta)X_B)\}\Big],
\end{aligned}$$

where w_A and w_B are the two agents' initial endowments, and losses are subtracted from the endowments. The objective of each agent is to maximize its own expected utility. Observe that

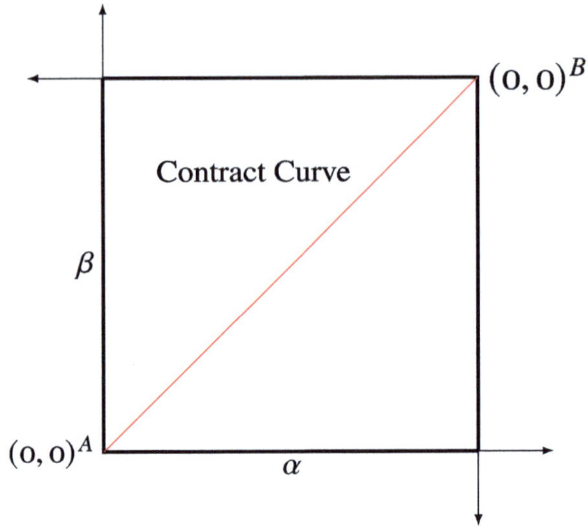

Fig. 3.11 Pareto efficiency for risk allocation

$$U^A(\alpha,\beta) \propto \frac{\gamma_A}{2}\left[\alpha^2\sigma_A^2 + \beta^2\sigma_B^2\right],$$
$$U^B(\alpha,\beta) \propto \frac{\gamma_B}{2}\left[(1-\alpha)^2\sigma_A^2 + (1-\beta)^2\sigma_B^2\right],$$

where the symbol $\propto$ means "is proportional to". Then the Pareto efficiency is achieved when

$$\text{MRS}^A = \text{MRS}^B,$$

which implies that the contract curve is given by

$$\frac{\alpha\sigma_A^2}{\beta\sigma_B^2} = \frac{(1-\alpha)\sigma_A^2}{(1-\beta)\sigma_B^2},$$

or equivalently, $\alpha = \beta$. In this case, the contract curve happens on the line that passes through (0, 0) and (1, 1), as shown in Fig. 3.11.

The result shows that any Pareto-efficient risk allocation rule in the context of proportional risk sharing is to split the aggregate risk $X_A + X_B$ equally between the two agents, regardless of each other's risk aversion rate γ_A and γ_B.

3.4.4 *Multi- to Single-Objective*

An alternative approach to finding Pareto efficiency is to turn the multi-objective Pareto optimization problem into a single-objective optimization problem. Let S be a set of possible solutions. Then Pareto optimality is defined as follows.

Definition 3.2 $x^* \in S$ is said to be a Pareto optimal solution if and only if there does not exist other $x \in S$ such that $f_i(x) \leq f_i(x^*)$ for $i = 1, 2, ..., m$ with strict inequality holding for at least one i.

We show in Appendix 3.A that, if all objectives are convex functions and S is a convex set, then the solution can be found through a single-objective optimization through the weighting method. The idea is to associate each objective function with a weighting coefficient and minimize the weighted sum of objectives. It is usually assumed that the weighting coefficients w_i are real numbers such that $w_i \geq 0$ for all $i = 1, 2, ..., m$ and the weights are normalized, i.e. $\sum_{i=1}^{m} w_i = 1$. Therefore, the equivalent weighting problem is stated as follows:

$$\text{minimize} \sum_{i=1}^{k} w_i f_i(x) \tag{3.2}$$
$$\text{subject to } x \in S,$$

where $w_i \geq 0$ for and $\sum_{i=1}^{m} w_i = 1$.

As shown in the previous section, Pareto optimal solutions are usually not unique. Equivalently, the weighting coefficients $\{w_i\}_{i=1}^{m}$ are arbitrary and each set corresponds to a particular Pareto optimal solution.

We elucidate the weighting method with some examples.

Example 3.8 Consider three objective functions f_1, f_2, and f_3 given by

$$f_1(x, y) = x^2 + (y - 1)^2$$
$$f_2(x, y) = (x - 1)^2 + y^2 + 2$$
$$f_3(x, y) = x^2 + (y + 1)^2 + 1.$$

We seek Pareto optimal solutions within the set $S = [-2, 2] \times [-2, 2]$. Then, the optimization becomes the weighting problem such that

$$\min f(x, y) := w_1 f_1 + w_2 f_2 + w_3 f_3,$$

where w_1, w_2, w_3 can be arbitrary weights as long as $w_1 + w_2 + w_3 = 1$ and $w_i > 0$ for $i = 1, 2, 3$. Then the first-order condition gives

$$\frac{\partial f}{\partial x} = 2w_1 + 2w_2(x - 1) + 2w_3 = 0,$$
$$\frac{\partial f}{\partial y} = 2w_1(y - 1) + 2w_2 y + 2w_3(y + 1) = 0.$$

The solutions are given by

$$x^* = w_2, y^* = w_1 - w_3.$$

For example, one set of weighting coefficients could be $w_1 = w_2 = w_3 = 1/3$, which yields the solution that $x^* = 1/3$, $y^* = 0$. Note f_1, f_2, and f_3 are all nonconstant convex polynomial functions, which achieve a local minimum only at the extreme points. Also note that axes of symmetry of all functions are located inside the region S. Thus, this is indeed a global minimum.

Example 3.9 Let us revisit the Pareto optimal solution to the risk allocation problem in Example 3.7. We can solve the same example using the weighting method. Recall that f is defined by

$$f(\alpha, \beta) := w_1 U^A(\alpha, \beta) + w_2 U^B(\alpha, \beta).$$

In order to maximize $f(\alpha, \beta)$, we obtain the first-order conditions

$$\frac{\mathrm{d}f}{\mathrm{d}\alpha} \propto w_1 \gamma_A \alpha - w_2 \gamma_B (1 - \alpha) = 0$$
$$\frac{\mathrm{d}f}{\mathrm{d}\beta} \propto w_1 \gamma_A \beta - w_2 \gamma_B (1 - \beta) = 0.$$

Thus,

$$\alpha = \beta = \frac{w_2 \gamma_B}{w_1 \gamma_A + w_2 \gamma_B}.$$

This is equivalent to solution from Example 3.7 that $\alpha = \beta$.

This chapter offers a light introduction to the economics of risk and insurance. A much more comprehensive review and in-depth analysis of insurance economics can be found in Eisen and Eckles (2011).

Appendix 3.A Equivalence of Pareto Optimality and Single-Objective Optimization

The solution of the weighting problem is always Pareto optimal given that weighting coefficients are all positive or if the solution is unique. If the optimization problem is convex, it is guaranteed that every Pareto solution is a solution to the weighting problem. Details can be found in Miettinen (2012) and Chankong and Haimes (2008).

Theorem 3.3 *The solution of the weighting problem* (3.2) *is Pareto optimal, provided that the weighting coefficients are positive, i.e.* $w_i > 0$ *for all* $i = 1, ..., m$.

Proof Let $x^* \in S$ be the solution of the weighting problem with positive weighting coefficients. Suppose that it is not Pareto optimal, i.e. there exists $x \in S$ such that $f_i(x) \leq f_i(x^*)$ for all $i = 1, ..., m$ and $f_j(x) < f_j(x^*)$ for at least one j. Since $w_i > 0$ for all $i = 1, ..., k$, it follows that $\sum_{i=1}^{m} w_i f_i(x) < \sum_{i=1}^{m} w_i f_i(x^*)$, a contradiction. □

We can also show that every Pareto optimal solution is also a solution to a weighting problem. To this end, we shall make use of the so-called generalized Gordan theorem, a known result from analysis. The proof of the generalized Gordan theorem can be found in Mangasarian (1994).

Theorem 3.4 (Generalized Gordan Theorem) *Let* f *be an m-dimensional convex vector function on the convex set* $\Gamma \subset \mathbb{R}^n$. *Then, either*

$$f(x) < 0 \text{ has a solution } x \in \Gamma,$$

or

$$pf(x) \geq 0 \text{ for all } x \in \Gamma \text{ for some } p \geq 0,\ p \in \mathbb{R}^m,$$

but never both.

It is rather straightforward to show that a Pareto optimal solution must also satisfy the single-objective optimization problem.

Theorem 3.5 *For convex multi-objective optimization, there exists a weighting vector for the Pareto solution* x^* *that solves a weighting problem.*

Now, we can prove the reverse direction.

Proof Let f be the m-dimensional convex vector function on the set $\Gamma \subset \mathbb{R}^n$ such that, for $x \in \Gamma$,

$$f(x) = (f_1(x) - f_1(x^*), ..., f_n(x) - f_n(x^*)).$$

Since x^* is Pareto optimal, there is no $x \in S$ such that

$$f_k(x) < f_k(x^*),\ f_j(x) \leq f_j(x^*),\ j = 1, ..., n,\ j \neq k.$$

It follows that there is no solution such that $f(x) < 0$. Thus, by the generalized Gordan Theorem we proved, there exists $p \in \mathbb{R}^n$ with $p \geq 0$ such that $\sum_{i=1}^n p_i(f_i(x) - f_i(x^*)) \geq 0$ for all $x \in S$. Let $p = \sum_{i=1}^n p_i$ and $w_i = p_i/p$. Note $w_i \geq 0$ and the sum of all w_i's is one. It follows that, for all $x \in S$,

$$\sum_{i=1}^{n} w_i f_i(x) \geq \sum_{i=1}^{n} w_i f_i(x^*),$$

which is the solution of the weighting problem.

Problems

1. Suppose that you are a contestant in the 10-briefcase "Deal or No Deal" game described in the thought experiment. If you are a student with no accumulation of wealth, what is the minimum offer of cash that you are willing to take instead of opening a case? Your utility function is assumed to be exponential with a risk aversion rate of $\alpha = 1$.
2. It is important to note that the utility function is usually not uniquely determined by the preference relation.

   a. Suppose that $u_1(x)$ is a utility function. Define $u_2(x) = au_1(x) + b$ for some real numbers $a > 0$ and b. Show that both utility functions would lead to exactly the same preference relation;
   b. Show that the marginal rates of substitution of the following three utility functions are the same:

   $$u_1(x, y) = 100xy^2 + 200, \quad u_2(x, y) = x^{1/3}y^{2/3}, \quad u_3(x, y) = \ln x + 2\ln y.$$

3. The linear utility function is an example of utility function with perfect substitutes. For example, if the person is indifferent between 1 apple and 2 bananas, then the apple is the perfect substitute for the banana with a ratio of 2. The opposite of that is the concept of the utility function with perfect complements. For example, say, each cup of coffee can only be enjoyed with two sugar cubes. If there are cups of coffee and two sugar cubes, then one cup is unenjoyable and redundant. In other words, the second cup of coffee without any additional sugar cube does not increase one's utility. The so-called Leontief utility function is an example of a utility function with perfect complements,

   $$U(x, y) = \min(ax, by).$$

   Find its marginal rate of substitution and draw its indifference curves.

4. Consider the risk allocation problem in Examples 3.7 and 3.9, where two economic agents enter into a risk sharing agreement. The means of losses from the two agents are given by $\mu_A, \mu_B \neq 0$.

   a. Using the marginal rates of substitution, show that the Pareto optimal solutions lie on the contract curve given by

$$\beta = \frac{\alpha\left(2\gamma_A\mu_B\sigma_A^2 - \gamma_A\gamma_B\sigma_A^2\sigma_B^2\right) - \gamma_A\mu_B\sigma_A^2 - \gamma_B\mu_A\gamma_B^2}{\gamma_B\sigma_B^2(2\mu_A + \gamma_A\sigma_A^2)}.$$

   Note when $\mu_1 = \mu_2 = 0$, $\beta = \alpha$, which is equivalent to Example 3.7.

   b. Show that the weighting method of single-objective optimization produces exactly the same solution.

5. Suppose that an individual is subject to some potential loss represented by $X \sim \text{Exp}(\lambda)$, i.e. the loss density function is given by

$$f(x, \lambda) = \begin{cases} \lambda e^{-\lambda x}, & x \geq 0 \\ 0, & x < 0 \end{cases}.$$

   The individual is willing to pay up to the amount P to reduce the variance of her risk. She considers buying either of two common types of insurance treaties, namely

   - Proportional insurance: $I(x) = \alpha x, 0 \leq \alpha \leq 1$;
   - Excess-of-loss insurance: $I(x) = (x - d)_+, d \geq 0$.

   In other words, an insurer can take on the risk $I(X)$ in exchange for a premium and the individual is left with the retained risk $X - I(X)$. Suppose that the premiums of both policies are both determined by the net premium principle, i.e.

$$\alpha\mathbb{E}[X] = \mathbb{E}[(X - d)_+] = P.$$

   Which type of insurance is better for this individual? Prove your claim.

6. Suppose that utility functions of consumers A and B are given by

$$U^A(x, y) = x + (y - 2)^2 + 3,$$
$$U^B(x, y) = (x - 3)^2 + 3y + 5.$$

   a. Identify the set of Pareto optimal solutions using the method of the marginal rate of substitution.

   b. Find the set of Pareto optimal solutions using the method of the weighting sum. Confirm if the two methods yield the same result.

7. Consider a risk allocation problem with three economic agents who are willing to share profits or losses with each other in order to improve their respective expected

utility. Suppose that agents A, B, and C have exponential utility functions such that

$$u_i(x) = \frac{1}{\gamma_i} e^{-\gamma_i x}, \qquad i = A,\ B,\ \text{and } C,$$

where γ_i is the risk aversion rate of agent $i = A, B,$ and C. Suppose that the three agents reached some profit/loss sharing agreement where each party carries a part of the total profit/loss. Agents A, B, and C carry the proportions α, β, and δ of the aggregate risk, respectively, where $\alpha + \beta + \delta = 1$. The profit/loss X_i of agent i is normally distributed with mean zero and variance $\sigma_i^2 > 0$ for all i. The three agents' expected utilities are given by

$$U^A(\alpha) = \frac{1}{\gamma_A}\mathbb{E}[\exp\{-\gamma_A(w_A - \alpha(X_A + X_B + X_C)\}]$$
$$U^B(\beta) = \frac{1}{\gamma_B}\mathbb{E}[\exp\{-\gamma_B(w_B - \beta(X_A + X_B + X_C)\}]$$
$$U^C(\delta) = \frac{1}{\gamma_C}\mathbb{E}[\exp\{-\gamma_C(w_C - \delta(X_A + X_B + X_C)\}],$$

where w_i are the agents' initial endowments, and losses are subtracted from the endowments. Show that the set of Pareto optimal solutions is given by

$$\gamma_A\alpha = \gamma_B\beta = \gamma_c\delta.$$

Thus, the allocation ratio should be inversely related to the risk aversion rate.

Chapter 4
Traditional Insurance

In this chapter, fundamental actuarial concepts in traditional insurance business models—including pricing, reserving, and capital management—are discussed. A variety of risk aggregation and capital allocation methods are addressed in contrast. Although the book is not intended to cover every facet of insurance models, it does provide the fundamental ideas and key components for risk management in traditional insurance businesses. Later chapters will use and expand on many of the same ideas in relation to decentralized insurance.

4.1 Pricing

Actuarial fairness refers to the principle that the premium paid by an insured should be commensurate with the risk taken by an insurer. The fairness principle enables insurers to charge differential premiums for policies of varying risk classes. In other words, in violation of fairness principles, an insurer would inevitably face the issue of adverse selection where overpriced low-risk individuals or entities exit the policies and underpriced high-risk individuals or entities persist, driving up the cost of insurance. Therefore, it is critical for insurers to develop sensible mechanisms that determine actuarial fairness.

This section introduces three different types of premium principles. It should be noted that, while the actual implementation in practice can be much more detailed and complex, these methods reflect the core pricing principles.

R. Feng, *Decentralized Insurance*, Springer Actuarial,
https://doi.org/10.1007/978-3-031-29559-1_4

4.1.1 Equivalence Premium Principle

A remarkable consequence of the LLN is that an insurer can charge a fixed premium to cover random claims. A very common form of payment for short-term insurance is a single premium to be paid at the inception of the contract. To consider the net cost of insurance liability, we often do not take into account expenses and taxes, and the single premium to be determined, denoted by P, is called a *net premium*. From the viewpoint of an insurance company, the *individual net liability* of the ith contract is the net present value of future liability less future income, denoted by L_i.

$$L_i = X_i - P. \tag{4.1}$$

Whenever there is no ambiguity, we shall suppress the subscript i for brevity. As common in the literature, we often refer to a positive net liability as a *loss* and the negative net liability as a *profit*.

Justified by the law of large numbers (see Theorem 3.1), we can set the premium such that the average claim outgo matches the average premium income, i.e. the average individual net liability would be zero

$$\mathbb{E}(L) = 0,$$

which implies, in the case of a single premium, that

$$P = \mathbb{E}(X),$$

where we use X as the generic random variable for X_i in an independent and identically distributed sample. We sometimes treat the pricing process as a mapping from the underlying random variable representing the underlying risk to a price, known as a premium principle, $\mathcal{P}$. The premium principle based on the law of large numbers is known as the *equivalence principle*, or net premium principle,

$$\mathcal{P}[X] = \mathbb{E}[X]. \tag{4.2}$$

The premium principle is frequently used in later chapters owing to its simplicity.

4.1.2 Portfolio Percentile Principle

While net premiums determined by the equivalence principle are used for many purposes in the insurance industry, a more common approach to setting premiums is based on profitability tests at the level of a business line. The *aggregate net liability*, denoted by $\mathfrak{L}$, is the sum of individual net liabilities in the whole portfolio of policies:

$$\mathfrak{L} = \sum_{i=1}^{n} L_i = \sum_{i=1}^{n} X_i - nP. \tag{4.3}$$

The most important objective of insurance risk management is to ensure that there is enough fund for the insurer to cover its liabilities. An alternative premium principle sets a premium P so that there is some minimum probability, say $\alpha \in (0, 1)$, that the aggregate net liability is negative (or the product is profitable). If $\mathfrak{L}$ is a continuous random variable,[1] then

$$\mathbb{P}(\mathfrak{L} < 0) = \alpha. \tag{4.4}$$

This is often called the *portfolio percentile premium principle*. Of course, applying such a principle requires one's knowledge of the aggregate net liability, the exact distribution of which can be difficult to obtain.

The central limit theorem (CLT) (Theorem 3.2) tells us that $\mathfrak{L}$ is approximately normally distributed with $\mathbb{E}(\mathfrak{L}) = n\mathbb{E}(L_i)$ and $\mathbb{V}(\mathfrak{L}) = n\mathbb{V}(L_i)$. Therefore, the premium P can be determined by

$$\mathbb{P}(\mathfrak{L} < 0) = \mathbb{P}\left(\frac{\mathfrak{L} - \mathbb{E}(\mathfrak{L})}{\sqrt{\mathbb{V}(\mathfrak{L})}} < \frac{-\mathbb{E}(\mathfrak{L})}{\sqrt{\mathbb{V}(\mathfrak{L})}}\right) \approx \Phi\left(\frac{-\mathbb{E}(\mathfrak{L})}{\sqrt{\mathbb{V}(\mathfrak{L})}}\right) = \alpha. \tag{4.5}$$

In view of (4.1) and (4.3), it follows immediately from (4.5) that the approximate portfolio percentile premium is given by

$$\mathcal{P}[X] = P = \frac{\sqrt{\mathbb{V}(X)}\Phi^{-1}(\alpha)}{\sqrt{n}} + \mathbb{E}(X). \tag{4.6}$$

It is clear that as n increases, the first term in (4.6) diminishes and the portfolio percentile principle converges to the equivalence premium principle.

4.1.3 Utility-Based Premium Principles

In insurance economics, utility theory is often used to explain the market price of insurance. Although such a pricing method is not directly used by pricing actuaries, it offers an approach to understanding the demand and supply of insurance and their impact on market price.

Consider that a risk-averse individual with utility u is willing to buy insurance against a risk X at a premium P. It is natural to ask what the maximum amount P is that he is willing to pay.

Note that the individual is only willing to buy the insurance if it improves his expected utility, i.e.

[1] Otherwise, we find P to be the smallest premium such that $\mathbb{P}(\mathfrak{L} < 0) > \alpha$.

$$\mathbb{E}[u(w - X)] \le u(w - P).$$

It is clear that the maximum premium P^+ is reached when

$$\mathbb{E}[u(w - X)] = u(w - P^+).$$

Hence, the maximum premium P^+ is determined by

$$P^+ = w - u^{-1} \left(\mathbb{E}[u(w - X)]\right). \tag{4.7}$$

Let μ and σ^2 be the mean and variance of X. We can use the following approximation of utility:

$$u(w - x) \approx u(w - \mu) + (x - \mu)u'(w - \mu) + \frac{1}{2}(x - \mu)^2 u''(w - \mu).$$

Therefore,

$$\mathbb{E}[u(w - X)] \approx u(w - \mu) + \frac{1}{2}\sigma^2 u''(w - \mu).$$

Similarly, we can approximate

$$u(w - P^+) \approx u(w - \mu) + (\mu - P^+)u'(w - \mu).$$

Recall that the maximum premium P^+ can be determined by

$$\mathbb{E}[u(w - X)] = u(w - P^+).$$

Therefore, the maximum premium P^+ is approximately

$$P^+ \approx \mu + \frac{1}{2}r(w - \mu)\sigma^2,$$

where r is known as the (absolute) risk aversion coefficient,

$$r(w) = -\frac{u''(w)}{u'(w)}.$$

If the insurer is also risk-averse with its own utility v, then the insurer is only willing to underwrite the risk if

$$\mathbb{E}[v(z + P - X)] \ge v(z),$$

where z is the insurer's initial capital. It is clear that the minimum premium P^- for the insurer must be given by

$$\mathbb{E}[v(z + P^- - X)] = v(z). \tag{4.8}$$

A deal can only be reached if one party wants to sell and the other party is willing to buy. Therefore, the insurance contract can only be sold if $P^+ \geq P^-$.

It is often assumed that an insurer is risk-neutral. In other words, the insurer is indifferent between a certain given outcome and a random outcome with the same expected value. This is because an insurer can benefit from risk diversification by the law of large numbers. Therefore, $\mathbb{E}[v(z + P^- - X)] = z + P^- - \mathbb{E}[X] = v(z)$, which implies that $P^- = \mathbb{E}[X]$. In other words, when an insurer is risk-neutral, the utility-based minimum premium agrees with the equivalence premium principle.

Example 4.1 Consider the demand for insurance from an individual whose exponential utility function is given by $u(x) = (1/2)(1 - e^{-2x})$. The initial financial position of the individual is assumed to be w. Suppose that the individual wishes to pass on to an insurer the risk X that follows a normal distribution with mean μ and variance σ^2. What is the maximum premium that the individual is willing to pay? Recall from (4.7) that

$$P^+ = w - u^{-1}(\mathbb{E}[u(w - X)]).$$

Observe that the inverse of the utility function is given by

$$u^{-1}(x) = -\frac{1}{2}\log(1 - 2x),$$

which means that

$$\mathbb{E}[u(w - X)] = \frac{1}{2} - \frac{1}{2}e^{-2w+2\mu+2\sigma^2}.$$

Thus, it follows immediately that the maximum insurance premium that the individual is willing to pay is given by

$$P^+ = \mu + \sigma^2.$$

Now consider the supply of the insurance from the viewpoint of an insurer. Suppose that the insurer is risk-neutral, i.e. its utility function is given by $v(x) = x$, and that the financial position of the insurer prior to entering into a new contract is given by z. What is the minimum premium at which the insurer is willing to underwrite the risk? It follows immediately from (4.7) that

$$P^- = v(z) - z + \mathbb{E}[X] = \mu.$$

It is clear from this exercise that the individual can pay more than the minimum premium that the insurer charges, i.e. $P^+ > P^-$. There is room for the two parties to negotiate and enter into an insurance contract.

Consider a different situation where the insurer also has an exponential utility function, say, $v(x) = 2(1 - e^{-(1/2)x})$. Note that in general an insurer as a financial institution has a higher risk tolerance and hence a smaller risk aversion rate than an individual. It follows from (4.7) that

$$P^- = \mu + \frac{\sigma^2}{4}.$$

Hence in this case $P^+ > P^-$. An insurance contract can be reached between an insurer and the insured.

The premium principles introduced in this book are merely a small sample of a variety of premium principles that have been studied in the academic literature. Detailed accounts of other premium principles and their properties can be found in Gerber (1979) and Denuit et al. (2006). These premium principles offer properties that make further analysis mathematically tractable in later chapters. However, one should bear in mind that these principles are in essence simplifications of much more complex pricing practices. Readers interested in actual ratemaking practices for property and casualty insurance are referred to Friedland (2021).

4.2 Reserve and Capital

As alluded to in Sect. 3.1.3, a key element of risk management by an insurer is to collect risks from policyholders and retain (at least a portion of) them for business. As the insurer takes on more risks, it is a standard practice and also required by regulators to set up reserve and capital to pay for its liability and prepare for adverse scenarios. In this section, we shall provide a brief overview of various types of reserve and capital.

4.2.1 Accounting

Like any business, an insurance company has a financial structure. A standardized tool to understand a company's financial structure is a balance sheet which summarizes its assets, liabilities, and net worth. A simplified example of a balance sheet is shown in Table 4.1.

In simple words, on one side of the balance sheet, *assets* include all resources the insurer owns. On the other side, *liabilities* reflect all the money the insurer owes to others. *Reserves* are what an insurer sets aside to provide for future claims. *Net worth* represents the excess of the value of an insurer's assets over the value of

Table 4.1 Balance sheet

Balance sheet (in thousands)			
Assets		**Liabilities**	
Investments		Unpaid losses	276,968
Bonds	519,107	Unpaid loss adjustment expenses	34,863
Stocks	130,274	Unearned premium reserves	233,853
Real estate	68,384	Ceded reinsurance payable	9,343
Cash and short-term investments	58,294	Other liabilities	73,773
Total investments	776,059	Total liabilities	628,800
Net premium receivable	134,743		
Reinsurance recoverable	15,322	**Net worth**	
Accrued investment income	9,683	Capital	253,492
Other assets	73,286	Surplus	126,801
Total assets	1,009,093	Total liabilities and net worth	1,009,093

its liabilities. In addition to reserve, insurers typically also earmark part of the net worth, often called *capital* to provide additional resources to cover losses. The key difference between reserve and capital is that reserve typically represents the expected insurance liability, whereas capital is often set aside to cover excessive losses beyond the reserve in adverse scenarios.

There are a few other accounting terms that are not critical to the discussion of the role of reserve and capital. For the sake of completeness, we briefly explain these concepts as well. Premiums receivable is an item for future premiums to be received; Reinsurance recoverable represents money that the reinsurers owe; Ceded reinsurance payable shows reinsurance premiums owed by the ceding company to the reinsurer; Unpaid losses refer to claims that are in the course of settlement. The term may also include claims that have been incurred but not reported. Unpaid loss adjustment expenses represent a cost insurance companies incur when investigating and settling an insurance claim. Unearned premium reserves refer to the amount of premiums written but not yet earned. Unearned premiums are portions of premiums collected in advance by insurance companies and subject to return if a client ends coverage before the term covered by the premium is completed.

4.2.2 Reserves

Example 3.4 offers a glimpse into concepts of reserve and capital. The reserve in that example is often referred to as net premium reserve, as it is determined on the basis of providing sufficient funding for future benefit payments. While net premium reserves provide a clear view of the basic principle, the determination of reserves can be far more complicated in practice as expenses, taxes, lapses, and other factors in actuarial assumptions are considered. In fact, different reserves are determined by financial reporting actuaries for various purposes and different audiences. Their differences arise from different accounting conventions.

1. Generally accepted accounting principles (GAAP) reserves: GAAP are a common set of standardized accounting principles, standards, and procedures that all companies follow when compiling financial statements. As the audience are investors, the primary objective is to provide an accurate reporting of the value of the business on a going-concern basis. Thus, "best estimates" of risk factors are used in actuarial assumptions.
2. Statutory accounting practices (SAP) reserves: An insurer must hold minimum reserves in accordance with statutory provisions of insurance laws. As the primary objective is to safeguard the insurer's solvency, SAP reserves are typically determined with conservatism on actuarial assumptions and accounting procedures.
3. Tax reserves: These liabilities are estimated for the purpose of determining income taxes.

4.2.3 Capital

Like reserves, different definitions of capital arise from different accounting conventions (e.g. GAAP, statutory, fair value, and economic), which prescribe different assumptions and methodologies of valuations applied to assets and liabilities. Accounting standards and procedures also vary greatly in different jurisdictions and are constantly evolving. There are many solvency/risk capital frameworks such as the U.S.'s NAIC Risk-Based Capital (RBC), Europe's Solvency II and Basel II, UK's Individual Capital Assessment (ICA), and Canada's OSFI Minimum Continuing Capital and Surplus Requirements (MCCSR). Since the main purpose of this book is to convey common principles, we shall refrain from getting into details of specific accounting conventions. Nonetheless, we shall summarize at a high level the general procedure of RBC requirement as an example of how a capital requirement is determined in practice.

Example 4.2 Risk-based capital (RBC) was established by the National Association of Insurance Commissioners (NAIC) in 1993 as an early warning system for US insurance regulation and has been updated on an annual basis. The RBC requirement specifies the minimum amount of capital an insurer is required to hold in order to

Table 4.2 RBC formula worksheet

Bonds (C-1o)			
Rating category	Book/Adjusted carrying value	Factor	RBC requirement
Long-term bonds			
Exempt obligations	———	× 0.000	= ———
Asset class 1	———	× 0.004	= ———
⋮			
Total long-term bonds	———		
Short-term bonds			
Exempt obligations	———	× 0.000	= ———
Asset class 1	———	× 0.004	= ———
⋮			
Total short-term bonds			———
Credit for hedging			———
⋮			
Total bonds			———

support its overall business operation in consideration of its size and risk profile. If an insurer does not meet the RBC requirement, then regulators have the legal authority to take preventive and corrective measures in order to protect policyholders and the stability of the insurance market.

Here, we only consider the life RBC formulas. For the most part of the current RBC model, the amount of capital required to be held for each item from the insurer's balance sheet, e.g. various assets, premiums, claim expenses, and reserve items, is calculated by applying various factors. In principle, the factor should be higher for items with greater risk and lower for those with less risk. The factors are predetermined and set by regulators. Table 4.2 gives an example of a worksheet for calculating RBC requirements for various bonds.

After the RBC requirements for all items are determined, they are grouped into the following five main categories of risks:

C-0 **Asset Risk—Affiliates**: The risk of assets' default for certain affiliated investments.

C-1 **Asset Risk—Other**: The risk of assets' default of principal and interest or fluctuation in fair value. This includes two subcategories: Asset Risk—Unaffiliated Common Stock and Affiliated Non-Insurance Stock (C-1cs) and Asset Risk—All Other (C-1o).

C-2 **Insurance Risk**: The risk of underestimating liabilities from business already written or inadequately pricing business to be written in the coming year.

C-3 **Interest Rate Risk, Health Credit Risk, and Market Risk**: The risk of losses due to changes in interest rate levels and the risk that health benefits prepaid to providers become the obligation of the health insurer once again, and risk of losses due to changes in market levels associated with variable products with guarantees. The three types of risks are assessed in subcategories C-3a, C-3b, and C-3c, respectively.

C-4 **Business Risk**: The risk of general business including mismanagement, lawsuits, fraud, regulatory changes, etc. It is assessed based on premium incomes, annuity considerations, and separate account liabilities. This category is further broken down into Premium and Liability (C-4a) and Health Administrative Expense (C-4b).

While RBC requirements for most risk categories are calculated by a factor-based approach, a notable exception is the C-3 risk, for which the NAIC implemented the RBC requirements in two phases. Implemented in 2001, Phase I addressed asset/liability mismatch risk for single premium life insurance and annuities including deferred and immediate annuities, guaranteed investment certificates, etc. Since 2003, NAIC has adopted several revisions for Phase II capital standards for variable annuities and other products with equity-related risks. Interested readers are referred to Feng (2018) for examples of RBC requirements for equity-linked insurance.

All RBC requirement data are then aggregated on a worksheet to determine the authorized control level RBC. It is assumed that each risk category is either completely independent of other risk categories or completely correlated with other risk categories. Using a statistical adjustment, known as "covariance adjustment", the authorized control level RBC is determined by

$$50\% \times \Big[\text{C-0} + \text{C-4a} + \sqrt{(\text{C-1o} + \text{C-3a})^2 + (\text{C-1cs} + \text{C-3c})^2 + (\text{C-2})^2 + (\text{C-3b})^2 + (\text{C-4b})^2}\Big]. \quad (4.9)$$

We shall offer some technical explanation of the formula in Example 4.3. The authorized control level of RBC is then compared with the Total Adjusted Capital, which is given by

$$\text{TAC} = \text{unassigned surplus} + \text{asset valuation reserve} + 50\% \times \text{dividend liability}.$$

Again all these items for the TAC calculation are determined by formulas prescribed by the NAIC. For more details, readers are encouraged to read an annual statement of a life insurance company based in the US, which is typically available in the public domain. Then the level of regulatory attention depends on the RBC ratio which is given by

$$\text{RBC ratio} = \frac{\text{Total adjusted capital}}{\text{Authorized control level of RBC}}.$$

Five levels of regulatory action may be triggered depending on the RBC ratio:

- None (RBC ratio > 200%) requires no action.
- Company action level (150% $\leq$ RBC ratio <200%) requires the company to prepare and submit an RBC Plan to the commissioner of the state of domicile. After review, the commissioner will notify the company if the plan is satisfactory.
- Regulatory action level (100% $\leq$ RBC ratio <150%) requires the insurer to submit an RBC Plan. After examination or analysis, the commissioner will issue an order specifying corrective actions to be taken.
- Authorized control level (70% $\leq$ RBC ratio <100%) authorizes the commissioner to take regulatory actions considered necessary to protect the best interests of the policyholders and creditors of the insurer.
- Mandatory control level (RBC ratio $\leq$ 70%) authorizes the commissioner to take actions necessary to place the company under regulatory control (i.e. rehabilitation or liquidation).

4.2.4 *Stochastic Reserving and Capital Requirements*

In contrast with the factor-based approach described in Example 4.2, there are certain types of risks, such as interest rate risk and equity risk, that can have severe financial implications and for which stochastic approaches are more suitable. In practice, the reserves and capital requirements for such risks are often determined by risk measures on financial consequences based on simulations. For example, solvency capital requirement "shall correspond to the Value-at-Risk of the basic own funds of an insurance or reinsurance undertaking subject to a confidence level of 99.5% over a one-year period" according to Solvency II, which is the regulatory framework for insurance companies in the European Union. In the next section, we offer an introduction to risk aggregation and capital allocation for traditional insurance businesses.

4.3 Risk Aggregation

Insurance businesses are built on the premise of pooling risks from tens of thousands of contracts and over different product lines. It is typical that insurers determine reserve and capital requirements at the contract level or for different lines of business in the first step. Thanks to risk diversification, it is unnecessary for insurers to hold up the sum of all reserves. Instead, lower aggregate reserve and capital requirements are determined at the corporate level. In the second step, the aggregate reserve and capital are allocated back to individual lines of business or even contracts. These two steps are commonly referred to as risk aggregation and capital allocation respectively,

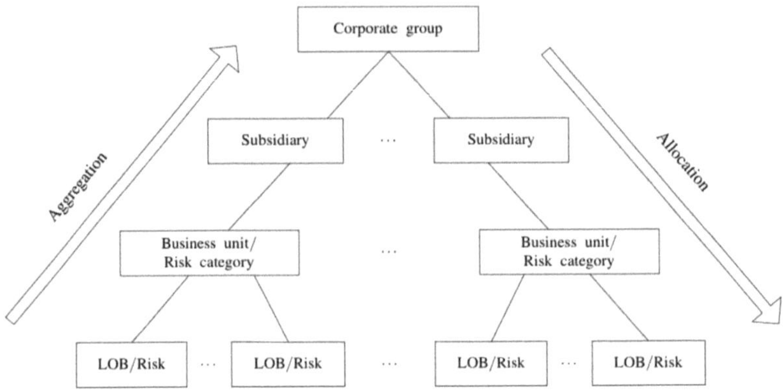

Fig. 4.1 Capital management hierarchy

which we shall discuss in this section and the next one. Figure 4.1 shows a typical structure of capital management in an insurance company.

4.3.1 An Illustration

Take the economic capital of a multi-line insurer as an example. Suppose that the insurer has three lines of business (LOB)—life, health, and property-casualty. There are specific methods for calculating economic capital for various lines of business. Readers are referred to Farr et al. (2008) for details.

Denote the standalone economic capital for the ith LOB by EC_i and the total economic capital by EC. Then it is often the case that

$$\sum_{i=1}^{n} EC_i > EC.$$

The difference is referred to as the diversification benefit. When the total economic capital is allocated to different lines of business, economic diversification is also distributed among them. The diversified economic capital for each LOB, which is the actual amount redistributed, can be determined by some proportional apportionment method

$$DEC_i = \frac{EC_i}{\sum_{j=1}^{n} EC_j} EC.$$

Such an apportionment method is one of many methods used in practice for risk aggregation and capital allocation. Figure 4.2 offers a visualization of its general principle.

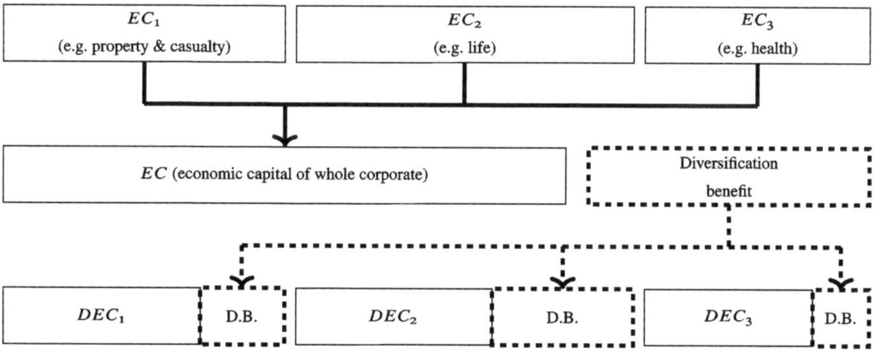

Fig. 4.2 Distribution of diversification benefit

4.3.2 *Variance-Covariance Approach*

The variance-covariance approach is commonly used in the banking and insurance industries. It is developed under the assumption that the potential losses (risks) from individual lines of business are normally distributed, and the minimal capital is determined by a risk measure of the aggregate risk which is proportional to its standard deviation. Suppose that there are n lines of business and their risk profiles (potential losses) are given by $(X_1, \cdots, X_n)$. The aggregate risk is thus determined by

$$S = \sum_{i=1}^{n} X_i.$$

This approach makes the assumption that all risks are normally distributed. For example, VaR can be used to determine minimal capital for business units with normally distributed losses:

$$\mathrm{VaR}_p(S) = \mathbb{E}[S] + \sigma(S)\Phi^{-1}(p), \qquad p \in (0, 1),$$

where $\sigma(S)$ is the standard deviation of the random variable S. Similarly, TVaR is also a popular choice of risk measure for capital management which has a similar structure:

$$\mathrm{TVaR}_p(S) = \mathbb{E}[S] + \sigma(S)\frac{\phi(\Phi^{-1}(p))}{1-p}, \qquad p \in (0, 1).$$

4.3.2.1 General Variance-Covariance Formula

Consider $\rho[S]$ to be any risk measure used to determine the risk capital at the corporate level, which has a particular structure $\rho[S] = \mathbb{E}[S] + c\sigma(S)$ for some constant c, which includes both value-at-risk and tail-value-at-risk. Recall that the variance of the sum of normal random variables can be written as

$$\sigma^2(S) = \sum_{i=1}^{n}\sum_{j=1}^{n} r_{ij}\sigma(X_i)\sigma(X_j),$$

where r_{ij} is the correlation coefficient of X_i and X_j. Then, the risk capital can be determined by

$$\rho(S) = \mathbb{E}[S] + \sqrt{\sum_{i=1}^{n}\sum_{j=1}^{n} r_{ij}\left(\rho(X_i) - \mathbb{E}[X_i]\right)\left(\rho(X_j) - \mathbb{E}[X_j]\right)}. \tag{4.10}$$

If the means of all losses are zero, it follows that

$$\rho(S) = \sqrt{\sum_{i=1}^{n}\sum_{j=1}^{n} r_{ij}\rho(X_i)\rho(X_j)}. \tag{4.11}$$

Furthermore, if all random variables $X_1, X_2, \cdots, X_n$ are mutually independent, then $r_{ij} = 0$ for $i \neq j$ in (4.11) and hence the formula simplifies to

$$\rho(S) = \sqrt{\rho(X_1)^2 + \cdots + \rho(X_n)^2}. \tag{4.12}$$

If $X_1, X_2, \cdots, X_n$ are perfectly correlated with each other, i.e. $r_{ij} = 1$ for all i, j, then

$$\rho(S) = \rho(X_1) + \cdots + \rho(X_n). \tag{4.13}$$

Example 4.3 Many capital regulatory frameworks, such as the NAIC risk-based capital, use an extended version of the variance-covariance approach to determine the total capital required for undertaking a variety of risks. Standalone capital is calculated for individual risk factors (e.g. C-0, C-1o, C-1cs, C-2, C-3a, C-3b, C-3c, C-4a, and C-4b). Then certain risk categories are grouped first and aggregated at the group level by applying the "variance adjustment" formula in (4.11). The purpose of such an adjustment is to take into account the potential diversification across different risks.

For example, the risks C-1o, C-1cs, C-2, C-3a, C-3b, C-3c, and C-4b are considered in a group, and it is assumed that the perfectly correlated pairs (C-1o, C-3a) and (C-1cs, C-3c) are independent of other mutually independent risks. The capital for aggregate risk in this group is given by the formula (4.12),

$$\sqrt{(\text{C-1o} + \text{C-3a})^2 + (\text{C-1cs} + \text{C-3c})^2 + (\text{C-2})^2 + (\text{C-3b})^2 + (\text{C-4b})^2}.$$

The risks C-0 and C-4a are considered mutually independent in another group and aggregated at the group level by the formula (4.13),

$$\text{C-0} + \text{C-4a}.$$

Last, the two groups are considered independent of each other and aggregated at the corporate level according to the formula for the authorized control level RBC.

Example 4.4 Suppose that a company faces both market and interest rate risks, from which the potential losses follow normal distributions. The following table gives the mean losses from the two types of risks and their covariance matrix. However, it is worth pointing out that the covariance matrix often has to be estimated from market data in practice. We can determine the company's capital requirement using $\text{VaR}_{0.95}$ as the risk measure and the variance-covariance approach.

Risks	Market	interest
Mean loss	10	20
Covariance matrix	$\begin{pmatrix} 3.5 & 1.8 \\ 1.8 & 5.6 \end{pmatrix}$	

Let X_1 and X_2 be the losses for market and interest rate risks respectively, and the aggregate risk $S = X_1 + X_2$. Note that $\Phi^{-1}(0.95) \simeq 1.64$. Then it is clear that S is normally distributed with mean 30 and variance 12.7. Thus,

$$\text{VaR}_{0.95}(S) = 30 + 1.64 \times \sqrt{12.7} \simeq 35.8445.$$

We can also determine the capital requirement via the variance-covariance formula. Note the correlation of the two risks is given by

$$r_{12} = \frac{1.8}{\sqrt{3.5 \times 5.6}} \simeq 0.4066.$$

Note that

$$\text{VaR}_{0.95}(X_1) = 10 + 1.64 \times \sqrt{3.5} \simeq 13.0682$$
$$\text{VaR}_{0.95}(X_2) = 20 + 1.64 \times \sqrt{5.6} \simeq 23.8809.$$

By the general variance-covariance formula, we obtain

$$\rho(S) = 30 + \sqrt{(13.0682 - 10)^2 + (23.8809 - 20)^2 + 2 \times 0.4066(13.0682 - 10)(23.8809 - 20)}$$
$$\simeq 35.4145.$$

The two results match with a minor difference due to some rounding error.

The variance-covariance approach has been studied and analyzed in the literature. Interested readers are referred to Arbenz et al. (2012), Dhaene and Vanduffel (2005), Filipovic (2009), and Pfeifer and Strassburger (2008) for detailed accounts of various applications in the banking and insurance industries.

4.4 Capital Allocation

Overall capital is often determined at the corporate level by the process of risk aggregation, as described in Sect. 4.3. Capital allocation refers to the process of dividing overall capital into more granular levels of the business. For example, to provide incentives for areas of potential growth, the management may develop a mechanism to allocate large capital to units with high risk and small capital to units with little uncertainty. For insurance companies, allocations may also be needed for reserves and other resources. In certain jurisdictions, it is a statutory requirement to allocate reserves to policies for tax purposes. The decision-making process behind allocation can be quite complex in practice, taking into account risk management policies, regulatory requirements, internal performance measures, etc. In this section, we provide an introduction to a few common approaches based on simple quantitative methods.

Definition 4.1 An *allocation principle* is a mapping from a random vector $(X_1, X_2, \cdots, X_n)$ to a real-valued vector $(K_1, K_2, \cdots, K_n)$ such that the sum $K_1 + K_2 + \cdots + K_n = K$ for a given total amount K.

Sometimes, the total amount of allocation is determined by a risk measure of the aggregate risk, i.e. $K = \rho(S)$ where $S = X_1 + \cdots + X_n$. A naive choice of allocation may be based on the risk measure of individual risk,

$$K_i = \rho(X_i).$$

However, such an allocation often lacks additivity. For example, if ρ is given by TVaR, then

$$K \leq K_1 + K_2 + \cdots + K_n.$$

To provide a remedy for additivity, we consider two sets of principles known in practice and in the literature.

4.4.1 Pro Rata Principles

It is a common practice to measure standalone risks by risk measures such as VaR and TVaR. In line with such a practice, a simple allocation method is to split the overall capital among individual risks in proportion to their standalone risk measures.

(*Allocation by proportion*) Given any risk measure ρ and the total amount K, the allocation to the ith risk is given by

$$K_i = \frac{\rho(X_i)}{\sum_{i=1}^n \rho(X_i)} K. \tag{4.14}$$

Haircut allocation principle: A common choice of risk measure in (4.14) for capital requirement is

$$\rho(X_i) = \text{VaR}_p(X_i).$$

While this principle is straightforward and easy to implement, a disadvantage is that it does not depend on the dependence among risks and hence ignores any diversification effect.

Quantile allocation principle: If the overall capital is set at $K = \text{VaR}_p(S)$, then the capital offers a level of security that it is sufficient to absorb losses with the probability p. However, with the haircut allocation principle, it is unclear at what level of security the allocated capital K_i offers to cover the ith risk. The quantile allocation principle is based on the idea of dividing capital to offer an equal level of security for all risks. Here, we describe the quantile allocation principle with the assumption that all risks are represented by continuous random variables, although it can be extended for all random variables with some technical nuances. The objective is to set the allocated amount for each risk by the same risk measure at the same probability level, i.e.

$$K_i = \text{VaR}_{\beta p}(X_i),$$

where β is to be determined so that $\sum_{i=1}^n K_i = K = \text{VaR}_p(S)$.

(*Allocation by marginal contribution*) Given any risk measure ρ and the total amount K, the allocation to the ith risk is given by

$$K_i = \frac{\rho_i^m}{\sum_{i=1}^n \rho_i^m} K,$$

where

$$\rho_i^m = \rho(S) - \rho(S - X_i).$$

We shall illustrate calculations under various capital allocation principles introduced earlier with a concrete example. Suppose that an insurer possesses a portfolio of n risks with which the potential losses follow a multi-dimensional normal distribution, i.e.

$$\boldsymbol{X} = (X_1, X_2, \cdots, X_N)^\top \sim N(\boldsymbol{\mu}, \boldsymbol{\Sigma}),$$

where we denote $\mu_i = \mathbb{E}[X_i], {\sigma_i}^2 = \mathbb{V}[X_i]$, and r_{ij} is the correlation coefficient of X_i and X_j, $\boldsymbol{\mu} = (\mu_1, \mu_2, \cdots, \mu_N)^\top$ and $\Sigma = (r_{ij}\sigma_i\sigma_j)_{i,j}$. We can determine allocated capitals under each allocation principle.

In the case of multivariate normal risks, we can compute the distribution of aggregate risk $S = \sum_{i=1}^n X_i = \boldsymbol{e}^\top \boldsymbol{X}$, where $\boldsymbol{e}$ is a column vector of ones with size n. We know from elementary probability class that any linear combination of normal random variables still follows normal distribution. Hence, it suffices to compute $\mathbb{E}[S]$ and $\mathbb{V}[S]$ to determine its distribution. Therefore, S is a normal random variable with

$$\mu_S := \mathbb{E}[S] = \mathbb{E}[\boldsymbol{e}^\top \boldsymbol{X}] = \boldsymbol{e}^\top \mathbb{E}[\boldsymbol{X}] = \boldsymbol{e}^\top \boldsymbol{\mu} = \sum_{i=1}^n \mu_i$$

$$\sigma_S^2 := \mathbb{V}[S] = \mathbb{V}[\boldsymbol{e}^\top \boldsymbol{X}] = \boldsymbol{e}^\top \mathbb{V}[\boldsymbol{X}]\boldsymbol{e} = \boldsymbol{e}^\top \boldsymbol{\Sigma} \boldsymbol{e} = \sum_{i=1}^n \sum_{j=1}^n r_{ij}\sigma_i\sigma_j.$$

Now consider the situation where the insurer is required to set aside a total capital based on VaR risk measure of its aggregate risk, i.e.

$$K = \mathrm{VaR}_p[S] = \mu_S + \sigma_S \Phi^{-1}(p).$$

The question is how to distribute the total capital to individual risks in an objective way. We show an allocation by proportion for illustration.

Example 4.5 (*Allocation by Proportion—Quantile Allocation Principle*) By definition, to guarantee an equal level of safety for all risks, there must exist a percentage β such that

$$K_i = VaR_{\beta p}[X_i] = \mu_i + \sigma_i \Phi^{-1}(\beta p) \text{ and } \sum_{i=1}^n K_i = K = \mu_S + \sigma_S \Phi^{-1}(p).$$

To solve for β, we note that

$$\Phi^{-1}(\beta p) = \frac{\sigma_S}{\sum_{i=1}^n \sigma_i} \Phi^{-1}(p),$$

which implies that

$$\beta = \frac{1}{p}\Phi\left(\frac{\sigma_S}{\sum_{i=1}^{n}\sigma_i}\Phi^{-1}(p)\right).$$

In general, as long as $p \geq 1/2$, since

$$\sigma_S^2 = \sum_{i=1}^{n}\sum_{j=1}^{n} r_{ij}\sigma_i\sigma_j \leq \left(\sum_{i=1}^{n}\sigma_i\right)^2 = \sum_{i=1}^{n}\sum_{j=1}^{n}\sigma_i\sigma_j,$$

we must have

$$\beta \leq \frac{1}{p}\Phi(\Phi^{-1}(p)) = 1,$$

which means that the level of security with the aggregate risk is always no less than that with individual risks. In other words, such an allocation principle takes into account the diversification effect among individual normal risks when $p \geq 1/2$. This is not surprising due to the fact that VaR is indeed subadditive for normal risks when $p \geq 1/2$. See (2.18) for the subadditivity of VaR for normal random variables. However, when $p < 1/2$, it is easy to see that

$$\beta > \frac{1}{p}\Phi(\Phi^{-1}(p)) = 1,$$

which means that the level of security with the aggregate risk is less than that with individual risks. In other words, under such an allocation principal, the capital requirement for the aggregate risk is higher than the sum of those for individual risks treated separately.

4.4.2 *Euler's Principle*

Euler's principle follows the same line of logic as the allocation by marginal contribution. The key difference is that the former measures the marginal contribution by the rate of change with the exposure to a particular risk, whereas the latter measures the marginal contribution by incremental change with the addition of a full unit of risk.

(*Euler allocation*) Given any positive homogeneous risk measure ρ for which $\rho[S + hX_i]$ is differentiable with respect to h for all $i = 1, \cdots, n$ and the total amount $K = \rho[S]$, the allocation to the ith risk is given by

$$K_i = \left.\frac{\partial \rho[S + hX_i]}{\partial h}\right|_{h=0}. \tag{4.15}$$

While Euler's principle is named after the famous mathematician Leonhard Euler (1707–1783), it is obvious that he did not work on such a capital allocation problem, as the insurance industry was still in a nascent stage in his lifetime and did not have sophisticated capital management. The principle carries his name as it results from Euler's theorem for homogeneous functions, which we shall briefly illustrate for the self-completeness of the book.

Suppose that ρ is positively homogeneous and $\rho(\lambda)$ is differentiable with respect to $\lambda \in \mathbb{R}$. Define a multivariate function

$$f(\boldsymbol{c}) = \rho(c_1 X_1 + c_2 X_2 + \cdots + c_n X_n),$$

where the vector $\boldsymbol{c} = (c_1, \cdots, c_n)$ represents number of units invested in n risks. It is clear that $f(\boldsymbol{e}) = \rho(S)$ where $\boldsymbol{e}$ is the vector of 1's of dimension n. Since ρ is positively homogeneous, then for any $\lambda > 0$,

$$f(\lambda \boldsymbol{c}) = \lambda f(\boldsymbol{c}).$$

Euler's theorem for homogeneous function implies that

$$f(\boldsymbol{c}) = \sum_{i=1}^{n} c_i \frac{\partial f(\boldsymbol{c})}{\partial c_i}. \tag{4.16}$$

Here, we provide a simple proof of (4.16). On the one hand, we apply positive homogeneity to show that

$$\frac{\mathrm{d}}{\mathrm{d}\lambda} f(\lambda \boldsymbol{c}) = \frac{\mathrm{d}}{\mathrm{d}\lambda} \lambda f(\boldsymbol{c}) = \rho[c_1 X_1 + c_2 X_2 + \cdots + c_n X_n]. \tag{4.17}$$

On the other hand, we use the differentiability of f and obtain

$$\frac{\mathrm{d}}{\mathrm{d}\lambda} f(\lambda \boldsymbol{c}) = \sum_{i=1}^{n} c_i \frac{\partial}{\partial(\lambda c_i)} f(\lambda \boldsymbol{c}) = \sum_{i=1}^{n} c_i \frac{\partial}{\partial c_i} f(\boldsymbol{c}). \tag{4.18}$$

Equating both (4.17) and (4.18) gives the identity (4.16). Setting $\boldsymbol{c} = \boldsymbol{e}$ gives Euler's allocation principle

$$\rho[S] = \sum_{i=1}^{n} \left.\frac{\partial \rho[X_1 + \cdots + c_i X_i + \cdots + X_n]}{\partial c_i}\right|_{c_i=1} = \sum_{i=1}^{n} \left.\frac{\partial \rho[S + hX_i]}{\partial h}\right|_{h=0}.$$

It may be difficult sometimes to determine an explicit expression for K_i in (4.15), in which case the finite difference approximation may be used:

$$\left.\frac{\partial \rho[S + hX_i]}{\partial h}\right|_{h=0} \approx \frac{\rho[S + hX_i] - \rho[S]}{h},$$

where h is a suitable choice of a small number. Keep in mind that risk measures are often estimated from Monte Carlo simulations. If h is too small, then the ratio may not be accurate due to sampling errors of estimators. If h is too large, then the finite difference may not be a good approximation of the derivative.

Example 4.6 (*Value-at-Risk*) It can be shown that for any continuous random variables X and S,

$$\frac{\partial}{\partial h}\text{VaR}_p[S + hX] = \mathbb{E}\Big[X|S + hX = \text{VaR}_p[S + hX]\Big]. \tag{4.19}$$

Suppose that the total amount of capital requirement is determined by

$$K = \text{VaR}_p[S],$$

for some $p \in (0, 1)$, and that all risks and the aggregate risk are continuous random variables.

Now we can use Euler's principle to allocate the total capital among all individual risks. Since VaR is positively homogeneous, then we can use (4.19) to determine the Euler allocation for the ith risk

$$K_i = \mathbb{E}[X_i|S = \text{VaR}_p[S]].$$

It is easy to see that

$$\sum_{i=1}^{n} K_i = \sum_{i=1}^{n} \mathbb{E}[X_i|S = \text{VaR}_p[S]] = \text{VaR}_p[S] = K.$$

Example 4.7 (*Tail-Value-at-Risk*) It can be shown that for any continuous random variables X and S,

$$\frac{\partial}{\partial h}\text{TVaR}_p[S + hX] = \mathbb{E}\Big[X|S + hX \geq \text{VaR}_p[S + hX]\Big]. \tag{4.20}$$

Suppose that the total amount of capital requirement is determined by

$$K = \text{TVaR}_p[S],$$

for some $p \in (0, 1)$, and that all risks and the aggregate risk are continuous random variables. In view of (4.20), we can determine the allocation for the ith risk given by

$$K_i = \mathbb{E}[X_i | S \geq \mathrm{VaR}_p[S]].$$

It is easy to see that

$$\sum_{i=1}^{n} K_i = \sum_{i=1}^{n} \mathbb{E}[X_i | S \geq \mathrm{VaR}_p[S]] = \mathbb{E}[S | S \geq \mathrm{VaR}_p[S]] = K.$$

Example 4.8 (*Euler Allocation Principle with VaR*) Under Euler allocation principle with Value-at-Risk as the underlying risk measure, we know that $K_i = \mathbb{E}[X_i | S = \mathrm{VaR}_p[S]]$. Recall from Sects. 2.4.1 and 2.4.3 that, for any two normal random variables X_1, X_2 with mean μ_1, μ_2, variance σ_1^2, σ_2^2, and correlation coefficient r, the distribution of X_1 conditioning on $X_2 = x_2$ is also normal with

$$X_1 | X_2 = x_2 \sim N\left(\mu_1 + r\frac{\sigma_1}{\sigma_2}(x_2 - \mu_2), {\sigma_1}^2(1 - r^2)\right).$$

Observe that $\mathbb{C}[X_i, S] = \mathbb{C}[X_i, \sum_j X_j] = \sum_j \mathbb{C}[X_i, X_j] = \sum_j r_{ij}\sigma_i\sigma_j$. It follows that

$$\begin{aligned} K_i &= \mathbb{E}[X_i | S = \mathrm{VaR}_p[S]] = \mathbb{E}[X_i] + \frac{\mathbb{C}[X_i, S]}{\mathbb{V}[S]}(\mathrm{VaR}_p[S] - \mathbb{E}[S]) \\ &= \mu_i + \frac{\sum_j r_{ij}\sigma_i\sigma_j}{\sqrt{\sum_{i,j} r_{ij}\sigma_i\sigma_j}}\Phi^{-1}(p). \end{aligned}$$

Also note that the total capital is indeed the sum of allocations to individual risks,

$$K = \sum_{i=1}^{n} K_i = \mathrm{VaR}_p[S] = \mu_s + \sigma_S \Phi^{-1}(p).$$

Example 4.9 (*Euler Allocation Principle with TVaR*) We can also apply the Euler allocation principle to capitals based on the TVaR risk measure:

$$
\begin{aligned}
K_i &= \mathbb{E}[X_i|S \geq \mathrm{VaR}_p[S]] = \frac{1}{1-p}\int_{\mathrm{VaR}_p[S]}^{+\infty} \mathbb{E}[X_i|S=s]f_S(s)\,\mathrm{d}s \\
&= \frac{1}{\mathbb{P}(S \geq \mathrm{VaR}_p[S])}\int_{\mathrm{VaR}_p[S]}^{+\infty}\left(\mu_i + \frac{\mathbb{C}[X_i, S]}{\sigma_S{}^2}(s-\mu_S)\right)f_S(s)ds \\
&= \mu_i + \frac{\mathbb{C}[X_i, S]}{\sigma_S{}^2}(\mathrm{TVaR}_p[S]-\mu_S) = \mu_i + \frac{\mathbb{C}[X_i, S]}{(1-p)\sigma_S}\phi(\Phi^{-1}(p)) \\
&= \mu_i + \frac{\sum_j r_{i,j}\sigma_i\sigma_j}{(1-p)\sqrt{\sum_{i,j} r_{i,j}\sigma_i\sigma_j}}\phi(\Phi^{-1}(p)).
\end{aligned}
$$

Note that in this case the total capital is also the sum of allocations,

$$
K = \sum_{i=1}^{n} K_i = \mu_S + \frac{\sigma_S}{1-p}\phi(\Phi^{-1}(p)) = \mathrm{TVaR}_p[S] = K.
$$

Euler's principle has also been used in the literature on many other risk measures, such as Return On Risk Adjusted Capital (RORAC) measure in Baione et al. (2021) and high moment risk measures in Gómez et al. (2022).

4.4.3 *Holistic Principle*

As discussed earlier, the classic approach to capital management is to treat risk aggregation and capital allocation sequentially in two separate steps. Figure 4.3 offers a summary of these two steps. In essence, the risk aggregation is done at the corporate level to assess the overall potential losses by scenarios. A certain risk measure is chosen to quantify the minimum capital required to fund the business lines. Once the aggregate risk capital is determined, various strategies are used to allocate the aggregate capital to different lines of business. A criticism of such an approach is the lack of consistency in the two steps of risk aggregation and capital allocation. The holistic approach discussed in this section is proposed as a remedy to this issue.

4.4.3.1 Classic Result

Value-at-risk (VaR) is a classic example to show the thought process of determining a risk capital by striking a balance between two competing goals of capital management and risk management.

On the one hand, a prudent risk manager may prefer as much capital as possible in order for the corporate to absorb severe losses under adverse economic conditions; on the other hand, it is undesirable to hold excessive capital as it is kept in liquid assets

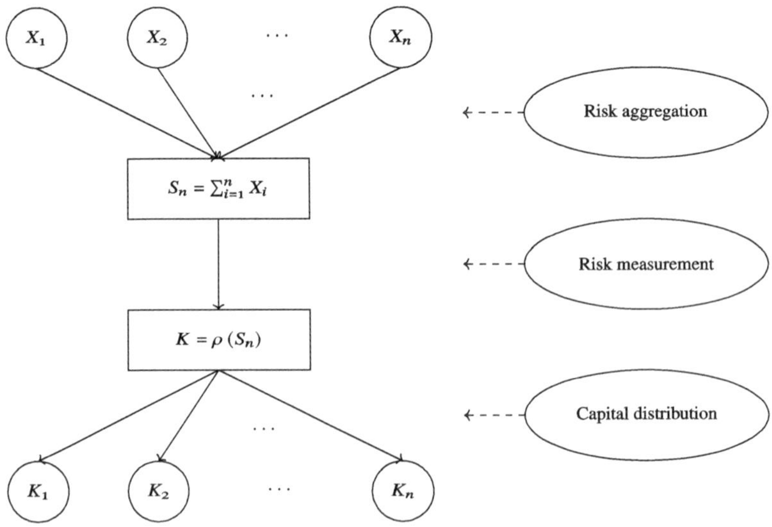

Fig. 4.3 Current practice

and has lower financial returns than other assets. Therefore, it is important to develop a sensible mechanism to strike a balance between competing capital management and risk management goals. In fact, such a delicate balancing act takes place implicitly when the VaR is used to determine the allocated capital.

The following optimization problem that minimizes the cost of capital and the expected shortfall has been discussed in Sect. 2.2.1:

$$\min\left\{ \underbrace{rK}_{\text{cost of capital}} + \underbrace{\mathbb{E}[(X-K)_+]}_{\text{expected shortfall}} \right\}.$$

When $0 < r < 1$, the optimal capital to the above-stated problem is given by $K^* = \text{VaR}_{1-r}(X)$. The first term is attributable to the capital management objective, as it represents the cost of capital at the borrowing interest rate r and penalizes excessive capital holding. The second term illustrates the risk management objective to minimize the expected shortfall beyond the allocated capital. The rate r also measures the relative importance of the capital management objective compared with the risk management objective. When $r = 0$, the capital management objective is ignored, and the decision is solely based on the risk management objective, which is to hold as capital the largest possible value of loss, i.e. $\text{VaR}_1(Y)$; when r rises, capital management exerts increasing influence on the capital setting, pushing down the required capital. When $r = 1$, the optimal capital becomes the smallest possible value of loss, i.e. $\text{VaR}_0(Y)$. That is, the rate r controls the trade-off between the excessive capital holding and the expected shortfall of loss, for the optimal capital.

	Capital management	Risk management	
1-st LOB	$\inf F_1\ (K_1)$	$\inf R_1\ (K_1; X_1)$	X_1
2-nd LOB	$\inf F_2\ (K_2)$	$\inf R_2\ (K_2; X_2)$	X_2
$\vdots$	$\vdots$	$\vdots$	$\vdots$
n-th LOB	$\inf F_n\ (K_n)$	$\inf R_n\ (K_n; X_n)$	X_n
Aggregation	$\inf F\ (\sum_{i=1}^n K_i)$	$\inf R\ (\sum_{i=1}^n K_i; S_n)$	S_n

Fig. 4.4 Pareto optimality

4.4.3.2 Competing Interest

In a business hierarchy, there could be competing interests both vertically and horizontally, with analogy to the VaR example. Vertically, individual lines of business (LOB) and the corporate may have their own goals. Horizontally, there are two conflicting management goals for each entity, namely capital management and risk management, that affect how capital is aggregated and allocated, as shown in Fig. 4.4. It is vital to first comprehend the requirements of each corporate entity.

On the one hand, the goal of capital management should be to increase capital efficiency, and capital should be kept to a minimum because capital is costly and is withheld from other corporate uses. Risk management, on the other hand, ensures that sufficient capital is available to absorb unexpected losses in the event of a disaster. To reduce the danger of insolvency, a risk manager may strive to accumulate as much capital as feasible.

Suppose that there are n lines of business in a corporate. The potential losses from these lines of business are denoted by $(X_1, \cdots, X_n)$, respectively. Hence, the aggregate loss at the corporate level is given by $S = \sum X_i$. We denote the capital management objective by $F(K)$. For example, the objective could be measured by the cost of capital rK for the amount of capital K and the borrowing interest rate r; or the goal could be measured by the square of a deviation from a previous capital level c, i.e. $(K - c)^2$. We shall denote the optimal capital by

$$K^F = \min F(K).$$

We denote the risk management objective by $R(S, K)$ depending on the aggregate risk S and the capital K. For example, the objective could be measured by the square of a capital mismatch $(S - K)^2$ or the capital shortfall $(X - K)_+$. The optimal capital denoted by K^R is given by

$$K^R = \min R(S, K).$$

Similarly, for the ith line of business, we could consider the capital management objective $F_i(K)$ under which the optimal capital is given by

$$K_i^F = \min F_i(K),$$

and the risk management objective $R_i(X_i, K)$ under which the optimal capital is given by

$$K_i^R = \min R_i(X_i, K).$$

4.4.3.3 Holistic Principle

The holistic principle for risk aggregation and capital allocation is proposed to strike a balance among these interests and priorities via Pareto optimality. As discussed in Sect. 3.4, Pareto optimality provides an efficient capital allocation such that it is impossible to improve any objective in the corporate, without comprising some other objectives for itself or its subsidiaries. Even though every component is allocated a sub-optimal capital, in the sense that it is not given by its optimal capital, the entire corporate achieves its best possible equilibrium in the sense of Pareto optimality. Using the weighting method described in Sect. 3.4, we can turn the Pareto optimality problem into a single-objective optimization.

Definition 4.2 The holistic principle for risk aggregation and capital allocation maps the random vector $(X_1, \cdots, X_n)$ to a constant vector $(K_1^H, \cdots, K_n^H)$, which is given by a minimizer of the following optimization problem:

$$\inf_{(K_1, K_2, \dots, K_n) \in \mathcal{K}} \sum_{i=1}^{n} \nu_i F_i(K_i) + \sum_{i=1}^{n} \omega_i R_i(K_i; X_i) + \nu F(K) + \omega R(K; S_n),$$

where $K = \sum_{i=1}^{n} K_i, \nu, \omega, \nu_i, \omega_i$ are non-negative deterministic weights, with at least one of them being positive for $i = 1, \cdots, n$, and $\mathcal{K}$ is a non-empty set of admissible capital allocations.

In the following, we consider a holistic principle with two types of considerations. Suppose that the insurer has a particular targeted reserve level, which could be the previous capital level c. Hence, the capital management goal is to avoid too far departure from some historic norm, i.e. $F(K) = (c - K)^2$. If the capital is too high, the excessive amount can be wasteful. If the capital is too low, it may not be sufficient to cover claims. Therefore, the risk management goal here is to minimize the magnitude of capital mismatch, i.e. $R(K, S) = \mathbb{E}[(S_n - K)^2]$. We can set up similar capital and risk management goals at individual lines of business, i.e. for the ith line of business,

$$F_i(K_i) = (c_i - K_i)^2, \qquad R_i(K_i) = \mathbb{E}\left[(X_i - K_i)^2\right],$$

where c_i is the targeted capital level for the particular line and K_i is the desired capital level. We consider all possible capital allocations without any further constraint, i.e. $\mathcal{K} = \mathbb{R}^n$. In summary, the holistic principle for capital allocation is given by

$$\inf_{(K_1,K_2,\dots,K_n)\in\mathbb{R}^n} \sum_{i=1}^{n} \nu_i \,(c_i - K_i)^2 + \sum_{i=1}^{n} \omega_i \mathbb{E}\left[(X_i - K_i)^2\right] + \nu\,(c - K)^2 + \omega\mathbb{E}\left[(S_n - K)^2\right].$$

Let us first consider the componentwise optimal capital levels. The corporate optimal capital for capital management purpose is given by

$$K^F = \arg\min(c - K)^2 = c.$$

The corporate optimal capital for risk management purposes can be determined by

$$K^R = \arg\min \mathbb{E}\left[(S_n - K)^2\right] = \mathbb{E}[S_n].$$

Similarly, individual LOB optimal capital for capital management is given by

$$K_i^F = \arg\min(c_i - K)^2 = c_i.$$

Individual LOB optimal capital for risk management is known to be

$$K_i^R = \arg\min \mathbb{E}\left[(X_i - K)^2\right] = \mathbb{E}[X_i].$$

It is clear that different functions of the corporate demand different optimal capital levels.

It turns out that the optimal holistic principle for capital allocation can be determined through a mixture of arithmetic and harmonic weighting.

Arithmetic weighting

The optimal capital at the corporate level without the consideration of its subsidiaries can be determined by

$$\begin{aligned}\overline{K} &= \underset{K\in\mathbb{R}}{\operatorname{argmin}} \;\; \nu\,(c - K)^2 + \omega\mathbb{E}\left[(S_n - K)^2\right] \\ &= \alpha \underbrace{K^F}_{\text{Cap Mgmt}} + \beta \underbrace{K^R}_{\text{Risk Mgmt}},\end{aligned}$$

where

$$\alpha = \frac{\nu}{\nu + \omega}, \qquad \beta = \frac{\omega}{\nu + \omega}.$$

It is clear that the corporate optimal capital is an arithmetic average of optimal capital levels for capital and risk management goals. Similarly, the LOB optimal capital can be given by an arithmetic average as well:

$$\begin{aligned}\overline{K}_i &= \underset{K_i \in \mathbb{R}}{\operatorname{argmin}} \quad \nu_i \left(c_i - K_i\right)^2 + \omega_i \mathbb{E}\left[\left(X_i - K_i\right)^2 h_i\left(X_i\right)\right] \\ &= \alpha_i \underbrace{K_i^F}_{\text{Cap Mgmt}} + \beta_i \underbrace{K_i^R}_{\text{Risk Mgmt}},\end{aligned}$$

where

$$\alpha_i = \frac{\nu_i}{\nu_i + \omega_i}, \qquad \beta_i = \frac{\omega_i}{\nu_i + \omega_i}.$$

Harmonic weighting

It is typical that the sum of LOB optimal capital levels is greater than the corporate optimal capital, as illustrated in Sect. 4.3.1. This is due to the fact that the corporate enjoys risk diversification, which occurs as profits from one LOB offsets losses from another one. Hence, we shall refer to the corporate optimal capital level as diversified capital, whereas the LOB standalone optimal capital levels are viewed as undiversified capital. In the holistic approach, we do not simply divide up the diversified capital, as it may not provide adequate capital to individual LOBs. Instead, we look for an equilibrium where the risk management or capital management metrics for neither the corporate nor any individual LOB can be improved without compromising those of others. The resulting holistic principle for capital allocation can be determined by a harmonic weight of diversified and undiversified optimal capital levels:

$$K^H = A \underbrace{\sum_{i=1}^{n} \overline{K}_i}_{\text{undiversified}} + (1 - A) \underbrace{\overline{K}}_{\text{diversified}},$$

where A is given by the harmonic weight

$$A = \frac{\frac{1}{\nu+\omega}}{\frac{1}{\nu+\omega} + \sum_{r=1}^{n} \frac{1}{\nu_r + \omega_r}}.$$

An alternative approach to represent the holistic aggregate capital is given by the corporate optimal capital increased by diversification amount, i.e.

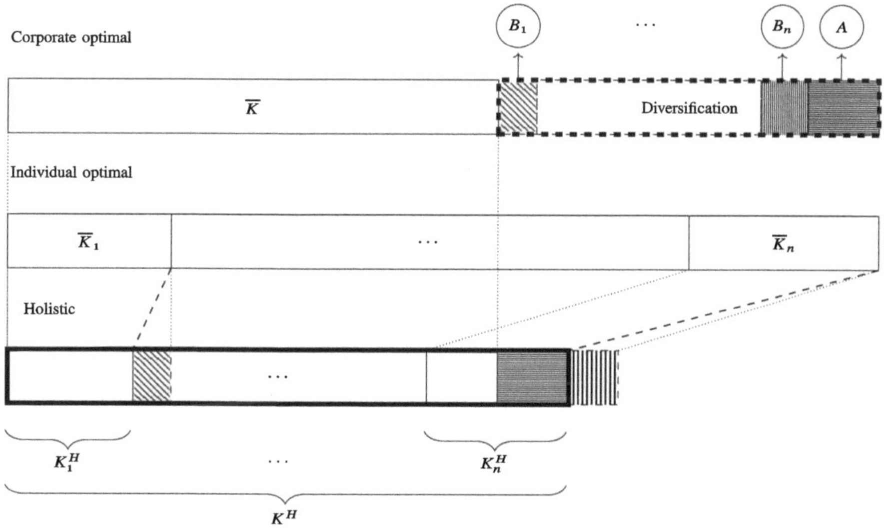

Fig. 4.5 Holistic aggregate and allocated capital levels—relationships with optimal aggregate and standalone capital levels

$$K^H = \overline{K} + A \underbrace{\left(\sum_{r=1}^{n} \overline{K}_r - \overline{K}\right)}_{\text{reduction by diversification}} .$$

Similarly, we can represent the holistic individual capital as

$$K_i^H = \overline{K} - B_i \left(\sum_{r=1}^{n} \overline{K}_r - \overline{K}\right), \tag{4.21}$$

where

$$B_i = \frac{\frac{1}{\nu_i + \omega_i}}{\frac{1}{\nu+\omega} + \sum_{r=1}^{n} \frac{1}{\nu_r + \omega_r}}, \qquad A + \sum_{i=1}^{n} B_i = 1.$$

In the current industry practice, the diversified capital is often distributed to each LOB/risk in proportion to its standalone capital, as shown in Fig. 4.2. Such a method is heuristic and lacks meaningful justification. The holistic principle for risk aggregation and capital allocation demonstrates a surprising similarity to the industry standard but is based on scientific reasoning. The connection of the holistic aggregate and allocated capitals, in relation to optimal standalone and aggregate capital levels for all LOBs and at the corporate level, is best illustrated in Fig. 4.5.

Decomposition of reduction by diversification benefit

The targeted and least squares holistic allocated capital is clearly a compromise made between optimal standalone capital for each LOB and the optimal aggregate capital at the corporate level. The capital allocation formula (4.21) shows that each LOB gives up some of its optimal standalone capital determined by a percentage B_i of the reduction by diversification. It is remarkable that the decomposition of reduced capital by diversification is based on *harmonic weighting*, since

$$A = \frac{\frac{1}{\nu+\omega}}{\frac{1}{\nu+\omega} + \sum_{r=1}^{n} \frac{1}{\nu_r+\omega_r}}, \quad B_i = \frac{\frac{1}{\nu_i+\omega_i}}{\frac{1}{\nu+\omega} + \sum_{r=1}^{n} \frac{1}{\nu_r+\omega_r}}, \quad A + \sum_{i=1}^{n} B_i = 1.$$

When a particular LOB/risk is viewed as significantly more important, i.e. the weight $\nu_i + \omega_i$ is comparatively larger, the weight B_i is smaller, and hence the LOB/risk is required to cut less from its optimal standalone capital.

Vertical balancing

In view of its competing interests with LOBs/risks, the corporate holds more capital than its optimal aggregate capital, and the increased amount is also based on the harmonic weighting, i.e.

$$K^H = (1 - A)\overline{K} + A \sum_{i=1}^{n} \overline{K}_i.$$

When the corporate has a higher priority for its optimality, i.e. the weight $\nu + \omega$ is larger, the weight A is smaller, and hence brings the holistic aggregate capital K^H closer to its optimal aggregate capital $\overline{K}$.

Early works on capital allocation principles based on optimization methods trace back to Dhaene et al. (2003, 2012). The holistic principle can be viewed as an extension of their work. Additional examples of the holistic principle can be found in Chong et al. (2021). The holistic principle can be applied beyond insurance applications. For example, spatial and temporal allocations of medical resources during a pandemic are studied in Chen et al. (2021). Capital management for cybersecurity risks has been proposed in Chong et al. (2022).

More discussions of risk aggregation and capital allocation can be found in Furman et al. (2021) and McNeil et al. (2015).

Appendix 4.A Proof of Holistic Principle

Recall that the holistic principle is determined by the optimization problem

$$\min_{\boldsymbol{K} \in \mathbb{R}^n} \sum_{i=1}^{n} \nu_i (c_i - K_i)^2 + \sum_{i=1}^{n} \omega_i \mathbb{E}\left[(X_i - K_i)^2\right] + \nu (c - K)^2 + \omega \mathbb{E}\left[(S_n - K)^2\right],$$

where $\boldsymbol{K} = (K_1, K_2, ..., K_n)$, $S_n = \sum_{i=1}^{n} X_i$, and $K = \sum_{i=1}^{n} K_i$. Then, by the first-order condition, we obtain

$$-2\nu_i(c_i - K_i) - 2\omega_i\,[\mathbb{E}(X_i) - K_i] - 2\nu(c - K) - 2\omega\,[\mathbb{E}(S_n) - K] = 0.$$

It follows that

$$[\nu_i + \omega_i + \nu + \omega]\,K_i + \sum_{j\neq i}^{n}(\nu + \omega)K_j = \nu_i c_i + \omega_i\mathbb{E}(X_i) + \nu c + \omega\mathbb{E}(S_n).$$

Thus, the optimal holistic capital for the ith line of business is given by

$$\begin{aligned} K_i = {} & A\left(\frac{\nu_i c_i + \omega_i\mathbb{E}(X_i)}{\nu_i + \omega_i}\right) + B_i\left(\frac{\nu c + \omega\mathbb{E}(S_n)}{\nu + \omega}\right) \\ & + \sum_{i\neq j}^{n} B_j\left[\left(\frac{\nu_i c_i + \omega_i\mathbb{E}(X_i)}{\nu_i + \omega_i}\right) - \left(\frac{\nu_j + \omega_j}{\nu_i + \omega_i}\right)\left(\frac{\nu_j c_j + \omega_j\mathbb{E}(X_j)}{\nu_j + \omega_j}\right)\right], \end{aligned}$$

where

$$A = \frac{1}{1 + \frac{\nu+\omega}{\nu_i+\omega_i} + \sum_{\gamma\neq i}^{n}\frac{\nu+\omega}{\nu_\gamma+\omega_\gamma}} = \frac{\frac{1}{\nu+\omega}}{\frac{1}{\nu+\omega} + \sum_{\gamma=1}^{n}\frac{1}{\nu_\gamma+\omega_\gamma}},$$

$$B_i = \frac{\frac{\nu+\omega}{\nu_i+\omega_i}}{1 + \frac{\nu+\omega}{\nu_i+\omega_i} + \sum_{\gamma\neq i}^{n}\frac{\nu+\omega}{\nu_\gamma+\omega_\gamma}} = \frac{\frac{1}{\nu_i+\omega_i}}{\frac{1}{\nu+\omega} + \sum_{\gamma=1}^{n}\frac{1}{\nu_\gamma+\omega_\gamma}}.$$

Note that

$$\frac{\nu_i c_i + \omega_i\mathbb{E}(X_i)}{\nu_i + \omega_i} = \frac{\nu_i c_i}{\nu_i + \omega_i} + \frac{\omega_i\mathbb{E}(X_i)}{\nu_i + \omega_i} = \alpha_i K_i^F + \beta_i K_i^R = \overline{K_i},$$

$$\frac{\nu c + \omega\mathbb{E}(S_n)}{\nu + \omega} = \frac{\nu c}{\nu + \omega} + \frac{\omega\mathbb{E}(S_n)}{\nu + \omega} = \alpha K^F + \beta K^R = \overline{K}.$$

Thus, the optimal allocated capital can be written as

$$K_i = A\overline{K_i} + B_i\overline{K} + \sum_{i\neq j}^{n} B_j\left(\overline{K_i} - \frac{\nu_j + \omega_j}{\nu_i + \omega_i}\overline{K_j}\right).$$

Observe that

$$A + \sum_{j=1}^{n} B_j = \frac{\frac{1}{\nu+\omega}}{\frac{1}{\nu+\omega} + \sum_{\gamma=1}^{n}\frac{1}{\nu_\gamma+\omega_\gamma}} + \sum_{j=1}^{n}\frac{\frac{1}{\nu_j+\omega_j}}{\frac{1}{\nu+\omega} + \sum_{\gamma=1}^{n}\frac{1}{\nu_\gamma+\omega_\gamma}} = \frac{\frac{1}{\nu+\omega} + \sum_{j=1}^{n}\frac{1}{\nu_j+\omega_j}}{\frac{1}{\nu+\omega} + \sum_{\gamma=1}^{n}\frac{1}{\nu_\gamma+\omega_\gamma}} = 1,$$

$$B_j \frac{\nu_j + \omega_j}{\nu_i + \omega_i} = \frac{\frac{1}{\nu_j+\omega_j}}{\frac{1}{\nu+\omega} + \sum_{\gamma=1}^{n} \frac{1}{\nu_\gamma+\omega_\gamma}} \cdot \frac{\nu_j + \omega_j}{\nu_i + \omega_i} = \frac{\frac{1}{\nu_i+\omega_i}}{\frac{1}{\nu+\omega} + \sum_{\gamma=1}^{n} \frac{1}{\nu_\gamma+\omega_\gamma}} = B_i.$$

This implies that

$$\begin{aligned} K_i &= \left(A + \sum_{j \neq i}^{n} B_j + B_i \right) \overline{K_i} - B_i \overline{K_i} + B_i \overline{K} - B_i \sum_{j \neq i}^{n} \overline{K_j} \\ &= \overline{K_i} + B_i \overline{K} - B_i \sum_{j \neq i}^{n} \overline{K_j} = \overline{K_i} - B_i \left(\sum_{j \neq i}^{n} \overline{K_j} - \overline{K} \right), \end{aligned}$$

and the holistic aggregate capital at the corporate level is given by

$$\begin{aligned} K = \sum_{i=1}^{n} K_i &= \sum_{i=1}^{n} \left[\overline{K_i} - B_i \left(\sum_{j \neq i}^{n} \overline{K_j} - \overline{K} \right) \right] \\ &= \left(1 - \sum_{i=1}^{n} B_i \right) \left(\sum_{j=1}^{n} \overline{K_j} \right) + \left(\sum_{i=1}^{n} B_i \right) \overline{K} \\ &= \overline{K} + A \left(\sum_{j=1}^{n} \overline{K_j} - \overline{K} \right). \end{aligned}$$

Problems

1. Suppose that an insurer deals with the risks $(X_1, \cdots, X_n)$ in its insurance portfolio. The risk X_i follows a normal distribution with the mean μ_i and the variance σ_i^2 for $i = 1, 2, ..., n$. One can determine the capital requirement for the insurer using the general variance-covariance formula with $\rho = \text{VaR}_p$.
   a. Determine the diversification benefit by comparing the sum of standalone economic capitals and the total capital;
   b. Show that the diversification benefit is always non-negative;
   c. There is no diversification benefit when the risks are perfectly correlated.

2. Assume that an insurer underwrites a portfolio of n lines of business with potential losses that follow a multi-dimensional normal distribution, i.e.

$$\boldsymbol{X} = (X_1, \cdots, X_N)^\top \sim N(\boldsymbol{\mu}, \boldsymbol{\Sigma}),$$

where $\boldsymbol{\mu} = (\mu_1, \cdots, \mu_N)^\top$ and $\boldsymbol{\Sigma} = (r_{ij}\sigma_i\sigma_j)_{i,j=1,\cdots,n}$ with mean μ_i and standard deviation σ_i for the ith line of business and correlation coefficient r_{ij} between i and jth lines. Determine the capital for the ith line using the allocation by marginal contribution and the following risk measure:

a. Value-at-risk;
b. Tail-value-at-risk;
c. Entropic risk measure.

3. Consider Euler's principle with the standard deviation as a risk measure. Set the total amount of capital K to be the standard deviation of the aggregate risk S, i.e. $\rho[S] = \sqrt{\mathbb{V}[S]}$. Determine the capital allocation for each risk in the insurer's portfolio.

   a. Let $f(\boldsymbol{c}) = \rho(c_1 X_1 + \cdots + c_n X_n)$ and $S_{\boldsymbol{c}} = c_1 X_1 + \cdots + c_n X_n$. Show that

$$\frac{\mathrm{d}f(\boldsymbol{c})}{\mathrm{d}c_i} = \frac{\mathbb{C}[X_i, S_{\boldsymbol{c}}]}{\sqrt{\mathbb{V}[S_{\boldsymbol{c}}]}}.$$

   b. Prove that the allocated capital for the ith risk by Euler's principle is given by

$$K_i = \frac{\mathbb{C}[X_i, S]}{\rho[S]}.$$

4. The holistic principle for capital allocation is an extension of the optimal capital allocation principles developed in the work of Dhaene et al. (2012). Consider the optimal allocation $\boldsymbol{K} = (K_1, K_2, \cdots, K_n)^\top$ with only risk management goal $R_i(X_i) = \mathbb{E}[\zeta_i(X_i - K_i)]$ and the admissible set $\mathcal{K} = \{\boldsymbol{K} \in \mathbb{R}^n | \boldsymbol{e}^\top \boldsymbol{K} = K\}$ for some fixed $K \in \mathbb{R}$ and $\boldsymbol{e}$ is a vector of ones. In other words, we look for $\boldsymbol{K} = (K_1, K_2, \cdots, K_n)$ such that

$$\min_{\boldsymbol{K} \in \mathcal{K}} \sum_{i=1}^{n} \mathbb{E}[\zeta_i (X_i - K_i)].$$

   Show that the optimal solution is given by

$$K_i = \mathbb{E}[\zeta_i X_i] + \left(K_i - \sum_{j=1}^{n} \mathbb{E}[\zeta_j X_j] \right).$$

Chapter 5
Decentralized Insurance

Traditional insurance is based on a centralized approach of risk transfer from the insureds to an insurer. A traditional insurance contract is a bilateral contract between an insured and an insurer. As the same contract is sold to millions of insureds, the insurer serves as a central service provider to all insureds. In contrast, decentralized insurance is organized in a community of participants where risks are transferred and shared with each other. A decentralized insurance scheme can be treated as a multi-lateral agreement among all participants. Instead of a central authority that sets the terms and conditions of financial arrangements, decentralized insurance is formed on the basis of mutual support. In this chapter, we introduce and examine the basic mechanisms for a range of decentralized insurance schemes developed in different parts of the world, including online mutual aid in China, peer-to-peer insurance in the West, takaful in the Middle East, and catastrophe risk pooling in Caribbean countries. To further illustrate the variety of innovations in the market, we shall also briefly touch upon other business models that can be broadly considered under the framework of decentralized insurance.

5.1 Background

Mutualization is a process that allows members of a community to bring together the resources and means of each for the benefit of all. The fundamental concept of mutualization for risk management has been practiced for centuries around the world where members in a society care for each other's financial needs in the event of misfortune. The roots of modern insurance date back to many early forms of risk sharing such as burial societies well documented in ancient Rome and Egypt, brotherly associations, frith, and religious guilds in Middle Ages Europe. The rise of modern insurance is an industrial revolution that transformed community-based

R. Feng, *Decentralized Insurance*, Springer Actuarial,
https://doi.org/10.1007/978-3-031-29559-1_5

mutual agreements for risk sharing into bilateral insurance contracts between an insurer and many policyholders that are protected by civil laws. While there is no direct cash payment between policyholders on the surface, insurers effectively play the role of intermediaries collecting payments from policyholders to pay for those who make legitimate claims, in a way similar to the early forms of community-based risk sharing.

Modern insurance is a centralized form of risk sharing, where an insurer serves as the central authority that provides insurance coverage and services to all policyholders. The standardization of insurance contracts enables insurers to sell them to tens of thousands of policyholders and scale up the risk pool, making the cost of insurance predictable and manageable. As an intermediary to facilitate the transfer of resources among policyholders, the insurer has a concentrated market power to price and design products, to set complex contract terms, and to decide to whom and when to pay claims. There is often mistrust by policyholders that insurers overprice products, fail to keep good on promises, and try everything to deny claims. The failure of an insurer can affect tens of thousands of policyholders, creating a rippling effect on the financial system. As such, the insurance industry is also tightly regulated to protect the interests of mass policyholders. While regulation is a solution to address the issues with centralized insurance, it also imposes heavy administrative costs to stay compliant with licensing, product regulation, market conduct, financial regulations, etc. The inhibiting cost increases entrance barriers for newcomers and reduces market competition, further exacerbating the centralization of the market with a small number of big insurers.

Decentralized insurance is a revival of the concept of mutualization with modern technology. A variety of innovative business models have emerged with the potential to disrupt the traditional insurance industry using telecommunication technologies such as smartphones, blockchain, and Internet of Things. These new tools could address some challenges facing the current insurance industry and improve efficiency in underwriting, risk pooling, and claims management. Along with this changing landscape of technology, the sharing economy, a socio-economic system built around the sharing of resources, is quickly evolving and gaining popularity in a wide range of industries. In contrast with the traditional server-client model in which a central authority provides services to all, the peer-to-peer nature of the sharing economy enables the exchange of goods and services among users without the heavy cost of intermediaries. The rise of the sharing economy brings a new channel for individuals to share their underutilized assets and receive financial rewards. Decentralized insurance is also gaining traction as the manifestation of the sharing economy in insurance coverage. By returning to a community-based mutual support model, decentralized insurance aims to reduce the role of a centralized insurer, or as many argue, "cutting out the middleman".

In this chapter, we consider a wide range of risk sharing schemes which have been observed in many parts of the world, including online mutual aid originating in China, peer-to-peer insurance in Europe and North America, takaful in the Middle East, and catastrophe risk pooling organized by many international organizations for African, Caribbean, and Central America countries, as well as various risk sharing schemes developed in the academic literature. *Online mutual aid* is a business model derived from crowdfunding, largely practiced in the area of critical illness coverage. In contrast with crowdfunding where donors pay but do not expect to receive a benefit in return, online mutual aid is a reciprocal agreement where participants collectively pay for benefits to those with legitimate claims in exchange for the same coverage for themselves. Unlike traditional insurance, online mutual aid does not require any ex-ante premium, and all payments are made ex post based on the total claims in each period. *Peer-to-peer insurance* arises when policyholders of traditional insurance, typically renter's or automobile insurance, participate in a pool of family and friends to share the costs of claims below their insurance deductibles. In other words, small claims below deductibles are paid out of a common fund to which each member contributes and large claims beyond the capacity of the policyholders' common fund are covered by traditional insurance policies. Any remaining surplus returns to participants. Such models are known to reduce the cost of insurance in comparison with traditional insurance with full coverage. *Takaful* is an Islamic alternative to traditional insurance developed by the West, as the latter is perceived as inconsistent with the Sharia law. Members make contributions to a takaful fund, which is managed by a financial institution as an operator and used to pay members' claims. When the fund is insufficient, an interest-free loan is provided by the operator to cover the rest of the claims. When the fund maintains a balance, the surplus is used to repay previous loans, and the rest is split between the takaful operator and members. Unlike a traditional insurer, a takaful operator does not own the takaful fund or take on the underlying risk. Takafuls have been developed for both property/casualty and life coverage. *Catastrophe risk pooling* is a financial arrangement among sovereign states to gain access to financial resources in the aftermath of catastrophes. While insurers can provide coverage for disaster payouts with some deductibles and policy limits, the participating countries also have to share some financial burdens with each other. Such a risk sharing arrangement among sovereign states bears close resemblance to community-based insurance strategies among individuals.

5.2 Online Mutual Aid

The online mutual aid model was developed in China around 2011 and derived from crowdfunding models. In fact, the very first online mutual aid platform was created to provide a crowdfunding channel for cancer patients. Instead of relying only on charity, the platform offered a way for participants, mostly cancer patients, to provide financial support to each other. Participants committed to sharing the cost of medical expenses up to some limit of those who went through expensive medical treatments. Each member is not only a donor to someone's causes but also can be a recipient of others' donations. In essence, mutual aid is a way of turning one-way charitable acts from one member to another into bilateral mutually beneficial financial agreements among peers. It was not until 2016–2018 that online mutual aid rose to public attention as large capitals from big tech firms started to pour into

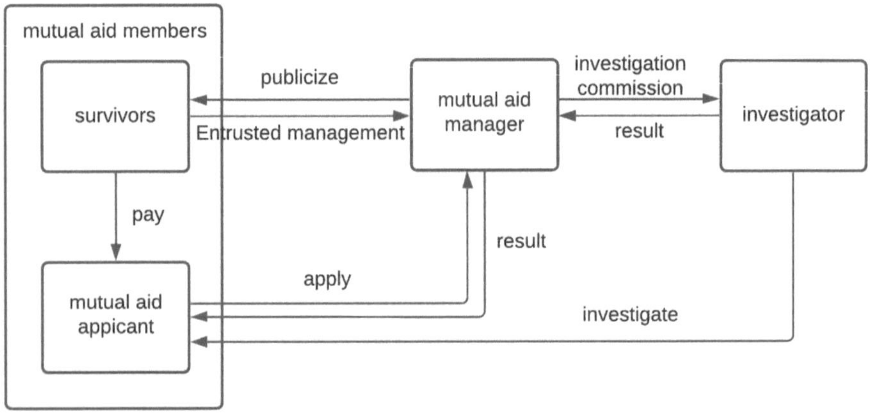

Fig. 5.1 Dynamics of mutual aid participation

the health insurance market. China's insurance industry was tightly regulated, and the regulator had not issued any insurance career licenses for many years. As the mutual aid model closely resembled mutual insurance, tech firms viewed this as an opportunity to enter the insurance industry. While its long-term success is yet to be seen, this innovative model provides a new tool for decentralized risk transfer and management.

As described in the white paper by Ant Group Research Institute (2020), the common practice of a mutual aid model involves three parties—mutual aid members, the platform, and a third-party claims investigation agency. Their interactions are shown in Fig. 5.1. Upon entering the platform, each member is required to disclose their health conditions and to sign a community agreement outlining their benefits and obligations. There is no underwriting procedure as with commercial health insurance. When a member is diagnosed with a covered critical illness, he/she can submit a claim for mutual aid. The platform calls in an investigation agency to verify the validity of claims in reference to the conditions stipulated in the agreement. Once a claim passes through the initial investigation, the platform publicizes the results to all mutual aid members. If there is any public dispute, a jury consisting of selected mutual aid members is formed to vote on an "accept/deny" decision. The public disclosure procedure is viewed by practitioners as an effective deterrence of fraudulent claims, which is not possible for traditional insurers due to regulations. It is typical that the platform levies an 8% management fee for each accepted claim, which is partly used to pay commissions to the investigation agency. The mutual aid model is largely a pay-as-you-go system. All accepted claims and associated management fees are aggregated at the end of each month and passed on to all members.

Online mutual aid is an innovative and collaborative coverage mechanism, different from traditional insurance in several ways. Firstly, online mutual aid is often much cheaper and financially more inclusive than health insurance. Unlike traditional insurance which often requires expensive premiums paid in advance, most mutual aid platforms require little to no fees to enter their platforms. The monthly shared costs among members vary by the actual number of claims but are often very affordable. The study by Chen et al. (2023) shows that the per-dollar coverage cost for comparable health insurance in China can be 500 times more costly than that of a mutual aid platform. The products have been quite popular with middle- to low-income people in small and medium-sized cities. The model fills a gap between the national health insurance program and the commercial health insurance industry. Secondly, online mutual aid promotes transparency and self-governance, which enables members to develop a sense of community and to provide the resources necessary to address members' financial needs. Unlike traditional insurance riddled with opaque operations, online mutual aid platforms offer public disclosures on claims and involve members in the decision-making process. Thirdly, online mutual aid platforms often do not have funding pools to avoid being perceived by the government as illegal fundraising. They often work with third-party banks to facilitate payments in members' escrow accounts. As mentioned earlier, the costs of all claims are shared among all members in each period. In contrast with traditional insurance, there is no necessity for reserves or risk capital. Hence, they are not regulated by the government. The reduction of compliance costs also contributes to the cost-effectiveness of such a funding mechanism.

To understand the underlying quantitative principles, we formulate the online mutual aid model with the following notation. For simplicity, we start with a homogeneous risk pool. Assume that there are a total of n members. Let I_i be the indicator for the claim status of member i, i.e. $I_i = 1$ if member i makes a claim in a given period; $I_i = 0$ otherwise. Denote by p the probability of loss for each member and by N the number of claimants out of all members in the period. The losses of all members are mutually independent. Hence, $N = \sum_{i=1}^{n} I_i$ follows a binomial distribution, i.e. $N \sim \mathsf{Binom}(n, p)$. Let b be a lump sum benefit payment to each member with a legitimate claim. Therefore, each member brings in his/her risk into the pool, i.e. $X_i = bI_i$. Suppose the platform imposes a fee on all claims at the percentage rate ρ. Therefore, the total cost of claims is given by $S = \sum_{i=1}^{n} X_i = bN$ and the total cost of claims and expenses is given by $(1 + \rho)S$.

There are several ways in which the total cost is allocated among members. Let Y_i be the cost of participation allocated to member i. Note that, in the case of homogeneous risks, all members are allocated the same cost. We do not distinguish payments by different individuals and hence drop the subscript when no ambiguity arises.

1. *All to claimants*: All members pay for the costs associated with claimants:

$$Y = \frac{(1 + \rho)S}{n}.$$

In the frictionless model, i.e. $\rho = 0$, we observe the conservation of losses,

$$\sum_{i=1}^{n} X_i = \sum_{i=1}^{n} Y_i.$$

2. *Survivors to claimants*: Only survivors, those who do not have legitimate claims, are asked to pay for the costs with claimants

$$Y = \frac{(1+\rho)S}{n-N}.$$

3. *Capped cost*: To avoid unlimited liability for participants, the platforms often assure them by imposing a maximum cost to pay each period. Let d be the cap on cost:

$$Y = \frac{(1+\rho)S}{n-N} \wedge d.$$

It is often the case that the number of N is very small in comparison with n in practice when the pool size is very large. Therefore, the distinction between "all-to-claimants" and "survivors-to-claimants" is negligible for practical purposes.

A detailed account of online mutual aid and issues regarding cost reduction, fairness, and adverse selection can be found in Chen et al. (2023).

5.3 Peer-to-Peer Insurance

The peer-to-peer insurance began its rise with the launch of an online platform in 2010 by a German start-up. Members form social networks, each with 10 members of family and friends with a common need for insurance coverage. They each contribute to a shared common fund. A portion of the fund is used to buy an excess-of-loss coverage from an insurer. Small claims are paid off the fund with the insurer stepping in to cover larger claims. When there are less claims than the total fund at the end of each period, the surplus would be returned to members. Overall, the members save more in the process than in any traditional insurance plan.

There are two kinds of peer-to-peer (P2P) business models, whose dynamics are shown in Figs. 5.2 and 5.3. In many papers such as Denuit and Robert (2021), they are called a broker model and a carrier model. While these names are reflective of how P2P insurance platforms operate from the perspective of regulators, we intend to focus on how they pass on risks to participating members or a third-party insurer. From the viewpoint of risk transfers, we will refer to them as group-covered or individually covered models in this paper. For a homogeneous risk group, the two models are in fact identical. Depending on allocation algorithms, the two models can be equivalent for heterogeneous risk groups.

In the group-covered P2P insurance model in Fig. 5.2, all participants pay their contributions into the same pool which is covered by a common insurer. The platform takes out a flat rate from the premium pool to cover their operating costs, and another specified proportion of the pool goes toward covering reinsurance. The distinction between a carrier platform or a broker platform depends on whether the platform

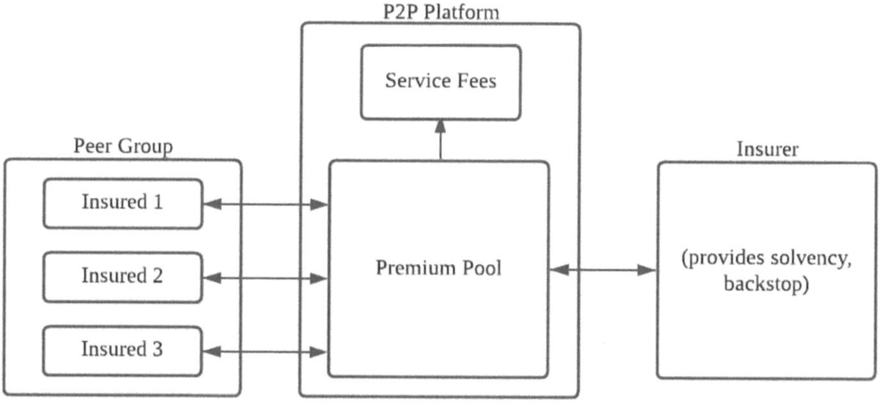

Fig. 5.2 Group-covered P2P model

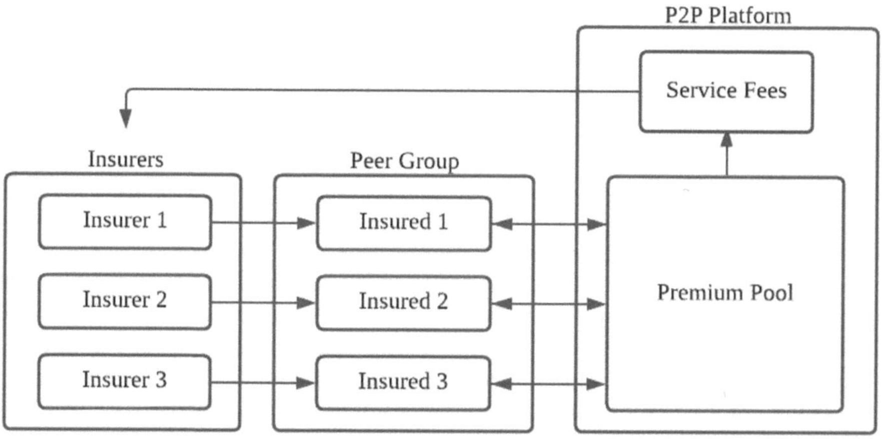

Fig. 5.3 Individually covered P2P model

steps into the role of an insurer to cover the excessive loss beyond the capacity of the premium pool.

In contrast, as shown in Fig. 5.3, some P2P platforms enable participants to cover each other's losses below the deductibles of their individual insurance policies. The main difference between the group-covered and individually covered P2P models is that members in the former collectively buy an insurance coverage for the funding pool, while members in the latter choose their own insurance policies to cover excessive losses. Note that, in Fig. 5.3, there could be multiple different insurers that cover their own insureds. There is only a single insurer for the entire pool in Fig. 5.2. Note that a P2P platform can serve as the insurer if it chooses a carrier model.

Despite these effectual changes in implementation, the P2P business model is based on one fundamental idea of sharing risk among peers. We outline the inner workings of this model in the following. Again we start with a homogeneous risk pool and show that the two business models are the same. For better comparison, we shall use as much the same notation as possible with other models. We shall denote by d the P2P contribution from each individual. Therefore, the capacity of the funding pool is given by nd. We also use $(x)_+$ and $(x)_-$ to denote the positive part and negative part of the real number x.

- Group-covered P2P insurance
  Assume that an insurer is willing to provide an excess-of-loss coverage to the funding pool at a gross margin θ. The insurance premium for the funding pool is given by $\Pi = (1+\theta)\mathbb{E}[(S-nd)_+]$, which is to be equally split among all members. As defined in the previous section, $S = \sum_{k=1}^{n} X_k = bN$. Observe that each member is entitled to a cash refund should there be any surplus, i.e. $(nd - S)_+/n$. Therefore, each member's net cost is given by

$$\frac{1}{n}\Pi + d - \frac{1}{n}(nd - S)_+. \tag{5.1}$$

  Consider the frictionless model where $\theta = 0$. Let $Y = S/n$. Then the shared cost in a P2P insurance scheme is given by

$$g^p(Y) = \mathbb{E}[(Y-d)_+] + d - (d-Y)_+ = \mathbb{E}[(Y-d)_+] + d \wedge Y. \tag{5.2}$$

- Individually covered P2P insurance
  Each member buys an insurance coverage with a deductible d in anticipation that the insurer will pay the amount by which the benefit b exceeds the deductible. Note that the group is assumed to be homogeneous. While we use member j as a representative, cash flows are represented by the same expressions for all members and we shall not use any subscript. Therefore, each pays $\pi = (1+\theta)\mathbb{E}[(bI_j - d)_+]$. The loss below deductible shall be transferred to the P2P insurance pool. In other words, each brings in the risk $bI_j \wedge d$. Therefore, the funding pool is responsible for total claims up to $S \wedge nd$. If there is a positive balance, the remainder $nd - S \wedge nd = (nd - S)_+$ is expected to be split equally among all members. Therefore, each member is expected to pay

$$\pi + d - \frac{1}{n}(nd - S)_+. \tag{5.3}$$

Observe that in the case of a homogeneous pool, the two models (5.1) and (5.3) are precisely the same. While only the case of homogeneous risks is discussed here, the group-covered P2P insurance model and the individually covered P2P insurance model can also be made equivalent for heterogeneous risks as we shall illustrate in Sect. 8.3.

5.4 Takaful

Traditional insurance developed in the West fails to meet the mark in Muslim communities as many Islamic scholars, those who study and interpret the Quran and its hadiths, have deemed it unacceptable within the confines of Islamic law, or Sharia law. In other words, traditional insurance is considered sinful, or haram, as noted by Malik et al. (2019). The inherent idea of insurance, providing those who experience a loss with compensation, has been practiced within Islamic communities for centuries as documented by Billah (1998), however, the current means by which traditional insurance goes about insuring their policyholders raises concern among Muslims.

There are three main sins outlined in Islam that traditional insurance violates.

1. *Gharar* refers to uncertainty and risk. Islam prohibits any financial transaction with a clear element of uncertainty. For example, one cannot sell fish which are still in the sea. Traditional insurance offers a contract that doesn't offer guaranteed benefits.
2. *Maysir* is the idea of gambling. If we define gambling as contracted payments for uncertain payout, there is a resemblance between traditional insurance and gambling. A policyholder may make regular payments for an extended period of time and receive no benefits at all in that time, putting them at a net loss while the insurer has made a net gain on their policies.
3. *Riba* is based on the concept of interest. In Islam, money is considered a means of exchange, and thus should hold no inherent value on its own. Interest enables those investing in interest-bearing assets to earn money with only money. This imbues money with value, treating it as an asset itself rather than a tool for exchange, contrary to Islamic beliefs. In traditional insurance, once the premium is paid to the insurer, the insurer invests in interest-bearing assets.

Because traditional insurance is not permissible under Islamic beliefs, takaful arose as an alternative to providing the same peace of mind and financial stability that traditional insurance does while still complying with Sharia Law. Takaful, often known as "Islamic insurance", is a mechanism for businesses and individuals to protect themselves from the financial consequences of unforeseen incidents. It is a covenant between a group of individuals who agree to jointly indemnify loss or harm from a fund to which they collectively give and is built on social solidarity and collaboration.

Takaful operates by grouping people together and setting up a fund to which each member contributes. To avoid conflation with insurers, the takaful company is referred to as the operator. When a loss is incurred, participants use the money in the fund to pay off their losses. This fund and all account activities are managed by the operator, e.g. investment activities and redistribution of the fund to participants. Where traditional insurance requires its insureds to pay premiums, takaful participants make contracted donations to the takaful fund. In the event that the takaful fund is not sufficient in order to pay off all the claims of the participants, the operator steps in to cover the deficit with an interest-free loan, or the qard hasan. Translated from Arabic, qard hasan means a benevolent loan. This loan comes from the company's shareholders. Although the participants are not legally obligated to pay back this loan, any surplus from future periods is used to pay back this loan to reimburse shareholders. Typically, the qard hasan involves another pot of funds kept by the shareholders and operators where the loan is taken from and paid back.

According to Archer et al. (2011), takaful circumvents the issue with *gharar*, as it is not a bilateral contract between an insurer and a participant, i.e. all the money that goes toward paying losses is not money paid out from the operators. Donations are made on the premise that nothing is expected in return, just like donating to a charity. These donations are then used to pay off any claims among the participants; the operator does not bear the uncertainty for the participant as each loss is split between each participant, thereby completing the risk sharing process. No promise is sold between the operator and participants.

The fact that the takaful fund is the property of all participants, not the operator, addresses the problem of *maysir*. A takaful operator simply manages the fund on behalf of the participants. The owners of the fund, however, are the participants; it is their money, as mentioned by Wahab et al. (2007). With that being the case, it is no longer gambling since the money is always belonging to the participants, and the operators do not stand to win should a loss not occur. The only compensation operators receive is for their management services, independent of any losses.

Takaful eliminates the issue of *riba* by investing in Sharia-complaint, or halal, assets. One such example of halal investment options are sukuk. Sukuk are effective bonds backing halal assets, offering fixed-income payments at regular intervals. In traditional bonds, people who purchase bonds receive interest for loaning their money, however, Ho (2016) explains that in sukuk, the only money someone receives in return is from the profit generated by the company taking their loans. Other kinds of sharia-compliant investments typically will not involve conventional financing, alcohol, non-halal food production, processing, or packaging, gambling, adult entertainment, tobacco, or weapons as described by Morgan (2020).

Any money remaining in the fund after claims and operating fees are paid is returned to participants. This works as an incentive for the group to collectively have less claims, thereby reducing moral hazards in a manner similar to what deductibles do in traditional insurance. Takaful funds use a sophisticated system to determine how much to return to participants. Thresholds are set in place to act as a smoothing mechanism, limiting the number of times an operator would have to pull money from the qard hasan. The common practice is to have five threshold levels, three for the takaful fund, and two for the qard hasan, as shown in Fig. 5.4. Should the fund fall below the *minimum funds level*, a loan from the qard hasan fund will bring the takaful fund back up subject to the *maximum limit* to the loan. When the takaful fund has enough capital to exceed the *investment trigger point*, operators can begin to invest. Once the money in the takaful fund reaches the *surplus trigger point*, the amount above that point will be distributed to takaful participants and the operator. The remainder would act as a reserve for future terms. When the qard hasan fund exceeds the *dividend trigger point*, any additional proceeds will be given back to

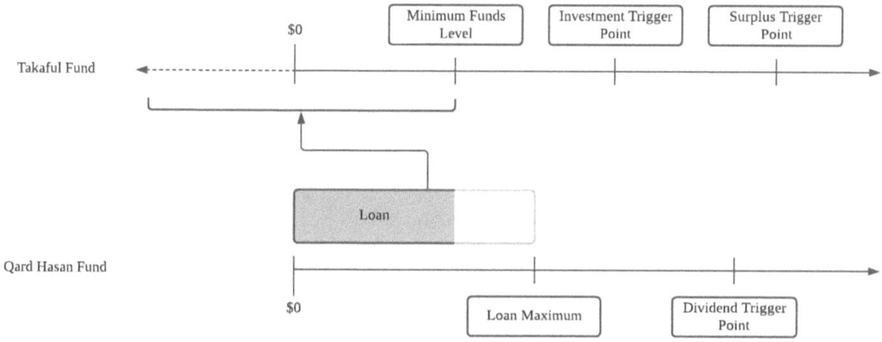

Fig. 5.4 Takaful and qard hasan funds

participants. This dividend payment can take many forms, such as cash rebates or discounted contributions for the next term.

To better understand the relationship of takaful with other decentralized insurance schemes, we take a minimalist approach to consider only the core elements to reflect on the main principles. Suppose that each takaful participant makes a contribution of the amount d to the takaful fund. Let R_t be the reserve after all payments have been made in the period proceeding time t. There is a total of claim payments S to be drawn from the takaful fund. Therefore, the net change in the fund in a given period is the total income less the total outgo, i.e. $nd - S$. Let h be the threshold for surplus distribution. In other words, any amount in excess of h at the end of each period shall be returned to members and/or takaful operators. Therefore, the takaful fund reserve satisfies the recursion relationship

$$R_t = h \wedge (R_{t-1} + nd - S)_+.$$

Observe that, when the balance $R_{t-1} + nd - S$ sinks below zero, the takaful operator is obligated to provide a qard hasan loan (interest-free loan) to save the fund from becoming insolvent. Let Q be the qard hasan loan amount, i.e.

$$Q = (R_{t-1} + nd - S)_-, \tag{5.4}$$

where $(x)_-$ is the negative part of real number x. Note that the total surplus of the takaful fund to be distributed is given by

$$(R_{t-1} + nd - S - h)_+.$$

There are three common takaful business models where the operator's fees are determined in different ways, namely the mudarabah model, the wakalah model, and the hybrid model.

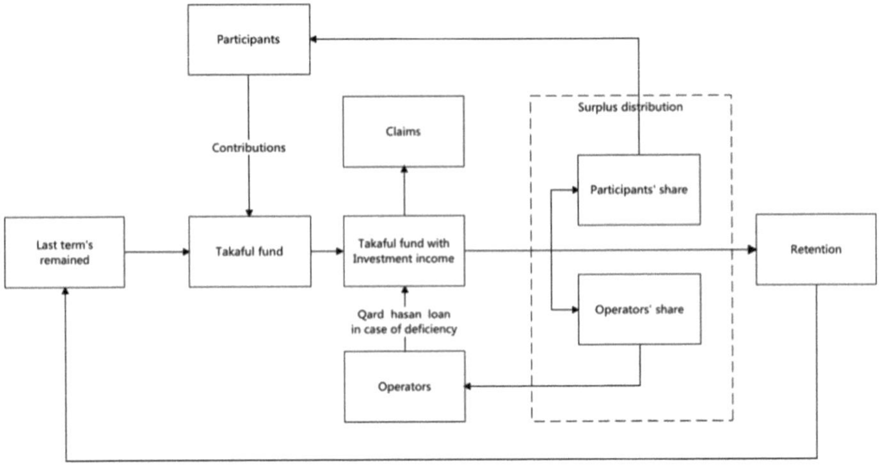

Fig. 5.5 Mudarabah Business Model

5.4.1 *Mudarabah*

The mudarabah model is also known as the profit-sharing model. Whenever there is a surplus, it is split between participants and operators in a pre-determined profit-sharing ratio. The underlying rationale is that the profit-sharing as opposed to interest is used to compensate the operator for its willingness to offer an interest-free loan whenever it is needed. The ratio can be anything the takaful operator and participants agree on. It is commonly a 70:30 split, with 70% going to the participants and 30% to the operators. Figure 5.5 allows one to follow this process from the beginning of the term to the end, then shows how the fund can be recycled for future periods. More details on the profit-sharing can be found in Saeed (2019).

To put it in mathematical terms, when there is a surplus, a proportion ρ^m of the surplus would be charged by the operator after the claim payment, i.e. $\rho^m(R_{t-1} + nd - S)_+$. It is typical that such a proportion is between 0 and 1. It should be noted that, when the participant's contribution is not enough to cover the claim amount, the qard hasan loan amount is provided as in (5.4). The reserve under the mudarabah model satisfies the recursion relationship

$$R_t = h \wedge (1 - \rho^m)(R_{t-1} + nd - S)_+.$$

The contribution of participants in t period is given by

$$nd - [(1 - \rho^m)(nd + R_{t-1} - S)_+ - h]_+.$$

It is worthwhile to point out that the takaful fund effectively provides a smoothing mechanism to reduce the fluctuation in the cost of participation over time. Observe that, had there been no smoothing mechanism as in the case of mutual aid, each participant is responsible for the cost $Y = S/n$. However, the takaful fund changes it to

$$d - \left[(1-\rho^m)\left(d + \frac{1}{n}R_{t-1} - Y\right)_+ - \frac{h}{n}\right]_+.$$

If we set the threshold for surplus distribution at $h = 0$, the reserve level is effectively fixed at zero for the takaful fund at the end of each period. In such a case, we can obtain a simplified form of the shared cost for each participant given by

$$g^m(Y) = d - (1-\rho^m)(d-Y)_+.$$

We shall use this cost as a representative for takaful models as a risk sharing mechanism. Note that $(d - Y)_+ = d - Y \wedge d$. Thus, the shared cost for each participant under the Mudarabah can also be written as

$$g^m(Y) = \rho^m d + (1-\rho^m)(Y \wedge d). \tag{5.5}$$

One concern of this model is that operator would only receive service payment if the fund makes a profit on investments at the end of the term. In the case of a deficit, the operator would not be able to gain any payment, and would instead have to supply the fund with the qard hasan. This is a concern not only to the operators but also to Sharia officials. The purpose of takaful is to provide participants with a means of financial security, however, this model introduces means that fall into a gray area in Sharia law. By sharing in the surplus at the end of the period, operators have an incentive to produce even more profit outside of simply providing financial support to their participants, once again discussed by Wahab et al. (2007). For this reason, different models have emerged in an effort to provide more transparent methods for providing coverage to participants while remaining within the constraints of Sharia law.

5.4.2 Wakalah

The wakalah model, also known as the agency model, makes payments to the operator based on the total contribution, as opposed to the surplus, in the mudarabah model. Notice in Fig. 5.6 that the operator's side of the flowchart (the top) will receive a payment earlier in the process than in Fig. 5.5, before claims were paid. Participants are entitled to their share of the surplus after claims are paid and reserve funds are retained for the next period. No matter the performance of the fund, the operators receive a fixed amount. It has been argued that such a model returns the priority for the fund to providing participants with coverage as opposed to earning profit for themselves. Details of the wakalah model can be found in Annuar et al. (2004).

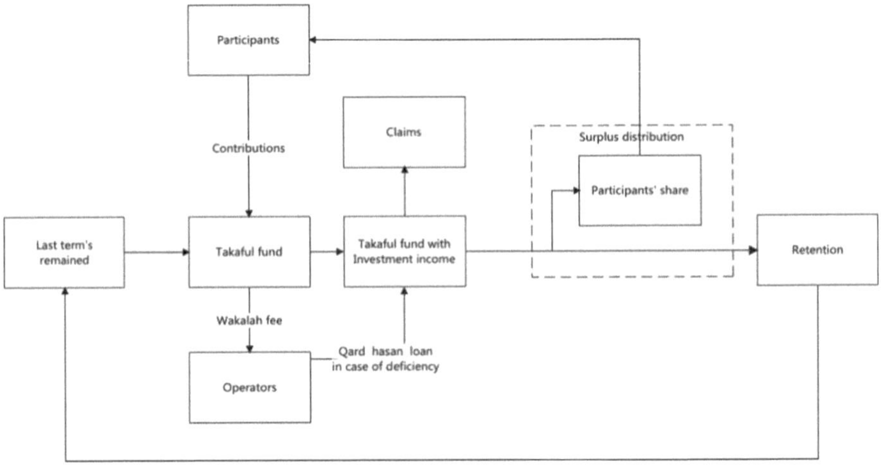

Fig. 5.6 Wakalah business model

We return to an algebraic representation of the risk sharing mechanism. The wakalah fee charged by the operator is a fixed percentage ρ^w of the total contributions, i.e. $\rho^w(nd + R_{t-1})$. The qard hasan loan amount is provided when the total contributions and the reserve are not enough to pay the claim, i.e. $Q = \left[(1-\rho^w)(R_{t-1}+nd) - S\right]_-$. The reserve under the wakalah model satisfies

$$R_t = h \wedge \left[(1-\rho^w)(R_{t-1}+nd) - S\right]_+ .$$

The contribution of participants in t period is given by

$$nd - [(1-\rho^w)(nd + R_{t-1}) - S - h]_+.$$

If we set the threshold for surplus distribution to be $h = 0$ and examine the risk sharing from the perspective of an individual, we arrive at a simplified case where the shared cost for each participant is given by

$$g^w(Y) = d - [(1-\rho^w)d - Y]_+.$$

Observe that $(x-y)_+ = x - x \wedge y$. Thus, the shared cost for each participant under the Wakalah model can also be represented by

$$g^w(Y) = (1-\rho^w)d \wedge Y + \rho^w d. \tag{5.6}$$

This expression gives an interpretation of the actual cost in two parts. The term $\rho^w d$ represents a fixed cost by Wakalah fee applied to the initial deposit. The term

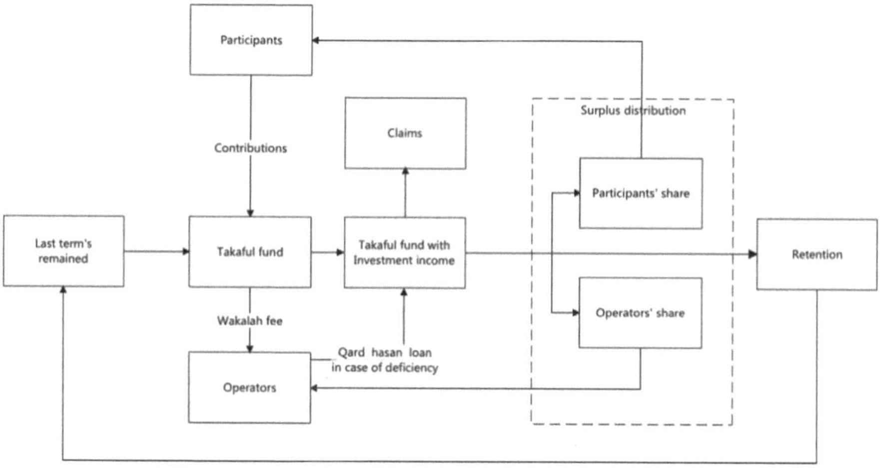

Fig. 5.7 Hybrid business model

$(1-\rho^2)d \wedge Y$ shows that the participant is liable for the shared cost Y up to the remaining balance of the deposit after the deduction of Wakalah fee, $(1-\rho^w)d$.

5.4.3 *Hybrid*

The hybrid model is a combination of the two prior business models which utilizes both the profit-sharing ratio, ρ^{hm}, and the wakalah fee, ρ^{hw}, i.e. $(R_{t-1}+nd)\rho^{hw} + \rho^{hm}((R_{t-1}+nd)(1-\rho^{hw}) - S)_+$. In Fig. 5.7, observe it is essentially a combination of the two previous diagrams.

The operator receives both the wakalah fixed fee and a share of the surplus on a mudarabah basis and receives the wakalah fee prior to any investments, then after investing activities and claim payments they receive a share of the investment income as well. The qard hasan loan Q will be provided when $(R_{t-1}+nd)(1-\rho^{hw}) < S$, and $Q = ((R_{t-1}+nd)(1-\rho^{hw}) - S)_-$. The reserve under the hybrid model satisfies

$$R_t = h \wedge [((R_{t-1}+nd)(1-\rho^{hw}) - S)(1-\rho^{hm})]_+.$$

The contribution of participants in t period is given by

$$nd - [((R_{t-1}+nd)(1-\rho^{hw}) - S)_+(1-\rho^{hm}) - h]_+.$$

Again, if we set $h = 0$, the simplified case of cost for each participant in the hybrid model is given by

$$g^h(Y) = d - [d(1 - \rho^{hw}) - Y]_+(1 - \rho^{hm}).$$

Compared to the mudarabah and wakalah operators, changes in investment performance will affect the income the operator receives, but not to the same degree as observed in the mudarabah model. There is also, of course, the fixed aspect similar to the wakalah model. The hybrid operator walks away with more income than both mudarabah and wakalah operators in this scenario.

5.4.4 Surplus Distribution

If there is a balance beyond the surplus trigger point in the takaful fund, the surplus is either returned to participants as cash back or used to reduce contribution amounts for the following term. There are generally three different methods for surplus distribution: pro rata, selective, and offsetting. The following descriptions of the three main surplus distribution methods are based on the outline provided by Soualhi (2016).

- Pro rata: Under the pro rata surplus distribution method, each participant has an equal right to the surplus, regardless of whether a claim was made in that term. In the case that all participants make equal contributions, each participant receives the same amount of money back. If the final surplus of the takaful fund with n participants is R_t, then a participant's share of the surplus is R_t/n.
- Selective: Under the selective surplus distribution method, only participants who did not make a claim, or survivors, would have the privilege of receiving a share of the surplus. The surplus is shared among survivors based on the proportion of each survivor who has contributed to the fund. It must consider (a) whether a participant is a survivor or a claimant, hence the indicator I_i in the numerator and (b) how many survivors share in the surplus. The number of survivors is simply the total number of participants minus the number of claimants in that period. Under the simplifying assumption of equal contributions, each survivor carries the equal share of cost: $R_t(1 - I_i)/(n - N)$.
- Offsetting: It is arguable that participants with more claims should be given less surplus in the following periods. With this method, all of the claims that have been made would be taken into account for eligibility for surplus distribution. Only when an individual's total contribution paid is larger than his/her total claim would he/she be eligible for surplus distribution. Suppose that participant i's net contribution to the fund at time t is $z_{it} = dt - b\sum_{k\leq t} I_{ik}$, where I_{ik} is the claims indicator for the ith participant at time k. At the end of each period, the amount of the surplus to be allocated to an individual i is calculated as $R_t(z_{it})_+/\sum_{j=1}^{n}(z_{jt})_+$.

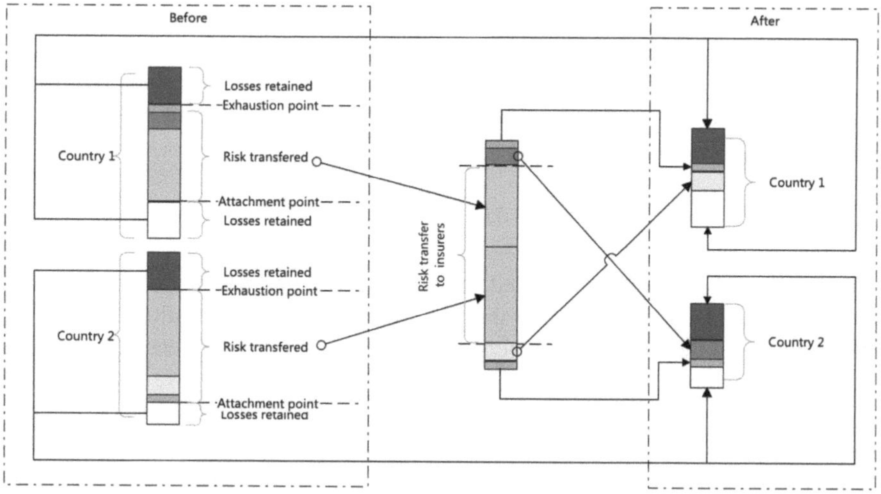

Fig. 5.8 CAT risk pooling

5.5 Catastrophe Risk Pooling

The world faces increased frequency and heightened severity of calamities due to extreme weather, and other natural and anthropogenic disasters. To alleviate the financial burdens of catastrophes on developing countries, there have been many developments of catastrophe, or CAT for short, risk pooling schemes at the international level facilitated by international organizations such as the world bank and other developed donor countries. Many best-known examples are the Caribbean Catastrophe Risk Insurance Facility (CCRIF), established in 2007, the African Risk Capacity (ARC), set up in 2012; and the Pacific Catastrophe Risk Assessment and Financing Initiative Facility (PCRAFI Facility), launched in 2016. While there are significant differences in their product designs and operational models, they share the fundamental principle that countries pool their risk-bearing capacities to collectively withstand severe losses beyond the capacity of an individual country. This section is devoted to the study of its mathematical formulation. Readers are referred to Bollmann and Wang (2019) for detailed analysis.

CAT risk pooling is typically set up as shown in Fig. 5.8. Suppose that there are n entities (typically countries) in the pool, facing risks (potential losses) due to particular perils denoted by $\{\tilde{X}_j, j = 1, \cdots, n\}$. As each entity has a natural risk-bearing capacity, it is typical that only excessive losses are passed on to the risk pool that can cause insolvency or liquidity crunch. The level below which the loss can be retained is known as the *attachment point*. The pool itself has a maximum funding capacity. Therefore, each entity can only transfer its risk to the pool by a certain maximum amount, known as the *coverage limit*. The amount above which losses cannot be covered by the pool is also referred to the *exhaustion point*, which is the sum of the attachment point and the coverage limit. It is also possible only a portion, known as ceding percentage, of the risk between attachment and exhaustion points is transferred to the pool. Therefore, the true risk from entity j to be transferred to the pool is determined by applying individual attachment point a_j, limit h_j, and ceding percentage β_j to the actual loss $\tilde{X}_j$:

$$X_j = \beta_j(h_j \wedge (\tilde{X}_j - a_j)_+). \tag{5.7}$$

In essence, the risk $\tilde{X}_j$ is divided into three layers, of which the bottom layer below the attachment point a_j and the top layer above the exhaustion $a_j + h_j$ are self-retained and the middle layer between a_j and $a_j + h_j$ is proportionally sent to the CAT risk pool.

The aggregate risk in the pool is given by $S = \sum_{j=1}^{n} X_j$. The CAT pool typically acquires a reinsurance contract on the aggregate loss S with attachment A and limit H. The choices of A and H have no connection to the individual attachment point, exhaustion point, or ceding percentage, (a_j, h_j, β_j) for $i = 1, \cdots, n$. In the absence of friction (overhead and profit margin), the cost of reinsurance would be

$$c = \mathbb{E}[H \wedge (S - A)_+].$$

Participants of the CAT pool have to decide on how to allocate the cost of reinsurance and retained aggregate risk. It is typical that the allocation of the retained aggregate loss to each participant can be proportional to the expected value of losses ceded to the pool, which is $\gamma_j = \mathbb{E}[X_j]/\mathbb{E}[S]$. The allocation of the cost of reinsurance c could be based on either of the following two percentages:

- Expected contributing losses to the pool $\alpha_j = \mathbb{E}[X_j]/\mathbb{E}[S]$.
- Conditional mean losses $\alpha_j = \mathbb{E}[X_j|S > A]/\mathbb{E}[S|S > A]$.

In other words, the entity j is financially liable for

$$\gamma_j[S \wedge A + (S - (A + H))_+] + \alpha_j c. \tag{5.8}$$

Consider the special case that the risk pool is homogeneous. In other words, all agents bring the same kind of risks into the risk pool. Then each agent shall be liable for an equal share of the aggregate risk. Hence, the allocation coefficient $\gamma_j = \alpha_j = 1/n$ for all $j = 1, \cdots, n$. Recall that in the absence of a third party, an insurer, each agent in the risk pool is responsible for an equal share of the aggregate risk $Y = S/n$. After a certain risk is transferred to the third party, each agent pays for

$$g^c(Y) = d \wedge Y + k(Y) + [Y - m]_+,$$

where $k(Y) = \mathbb{E}[(Y \wedge m - d)_+]$, $d = A/n$, and $m = (A + H)/n$. Therefore, each peer's liability Y is divided into three layers. The layer between d and m is transferred to an insurer and layers below d and above m are retained by the agent.

As the purpose of this chapter is to offer examples of decentralized insurance on a comparable basis, we have limited the discussion of the shared cost for the CAT risk pooling to a homogeneous group. We shall continue the discussion of CAT risk pooling for a heterogeneous group in a general framework of decentralized insurance in Sect. 8.4.4.

5.6 Other Decentralized Models

The practices of decentralized insurance have a long history in many forms around the world. The earlier sections merely offer a small sample of various practices, whose quantitative details are drawn from Feng et al. (2022b). There are also many other prevalent decentralized insurance models, some of which we shall briefly discuss without diving into details. These business models are decentralized in the sense that the risks are spread among members rather than borne entirely by a central entity.

5.6.1 Health Share

Health share plans are offered by faith-based non-profit organizations to pool funds from members to pay for each other's medical bills. Initially established by Amish and Mennonite communities, health share caring ministries offer financial support to their members for medical expenditures. The passage of the Affordable Care Act (ACA) in 2010 catalyzed the growth of health share plans in the US. The ACA individual mandate requires most Americans to have health insurance or pay a tax penalty, unless they qualify for exemptions such as health share plans. Therefore, healthcare plans marketed themselves as more affordable alternatives to health insurance. According to the Alliance of Health Care Sharing Ministries, there are 108 ministries with over 865 thousand members by 2021.

All members of the health share plan pay a monthly premium, part of which is set aside in a pool of funding. When a member receives a bill for covered medical services, he or she pays the bill themselves and then submits the bill to the health share organization for reimbursement. Once approved, the member receives compensation for the covered portion of the bill. While the compensation is paid out of the funding pool, the cost is in essence borne by all members contributing to the pool. Such a cost sharing process is very similar to that of P2P insurance. As health share plans are not considered as insurance, they are not subject to regulatory oversight to monitor their financial conditions. Health Share plans are typically designed for larger unexpected medical needs such as hospitalization for accidents. They commonly exclude services that are considered immoral by Christian beliefs such as treatment for maternity care for unmarried women, illness caused by unhealthy lifestyle choices, etc. It is argued that the low-premium and low coverage feature of healthcare plans contributes to an anti-selection. Health share plans appeal to younger and healthier people who seek exemption from ACA individual mandate while sicker people tend to stick with health insurance plans on the ACA marketplace. For the reasons above, health share plans are usually much more affordable than traditional health insurance. More details on health share plans can be found in Pistilli (2021).

5.6.2 *Multinational Pooling*

Multinational pooling is a mechanism developed in the insurance industry for decades to share risks among their clients' employee benefits plans. It allows multinational corporations to develop a virtual pool to share loss or profit among their subsidiaries within a corporation or among head offices of different corporations. Contracts included in such pools cover retirement, group life and survivors' benefits, disability, medical, accidental death, and critical illness. The insurer performs local accounting on each employee benefits plan in the pool and consolidates the data from all contracts to produce an overall profit and loss account. By balancing the losses from one country with the gains from another, the pool enables corporations to achieve greater stability of plan costs over time which is difficult to achieve by each subsidiary or corporate alone. It also allows participants to track the development of their employee benefits plans and exchange data with each other for transparency and governance. More details can be found in SwissLife (2018).

5.6.3 *Tontines*

The history of tontines predates the modern pension and retirement industry. It was a popular approach used by medieval European governments in the eighteenth and nineteenth centuries to raise capital often to fund wars against each other. The tontines are organized as a form of government-backed bonds. Each individual contributes a large lump sum to a common pool of funding managed by the issuing government. The proceeds from the bond are split on an annual basis among the survivors in the pool. As time goes by, there are fewer participants remaining alive and hence fewer people to split the proceeds. Such a scheme is appealing to investors because the deceased members' shares are transferred to survivors and all investors bet on their longevity. There was a well-known story where a French widow, Charlotte Bonnemay, lived to age 96, who at the time of her death enjoyed an annual tontine payment of 73,500 livres from a subscription of 300 livres at the inception.

Modern annuity is a centralized arrangement where each policyholder acquires a guaranteed series of payments until death by paying a lump sum to an insurer. If a policyholder exceeds the life expectancy, the insurer continues to pay at a loss. The insurer has to bear the financial responsibilities of all its policyholders. In contrast, tontine is a decentralized version of annuity. The issuing authority itself does not take on any risk in facilitating tontine pools. The shares from the deceased members are directly paid to the survivors. In other words, all longevity risks are spread among the tontine investors. Readers can find interesting stories and the history of tontine and arguments for its revival in the twenty-first century in Milevsky (2015).

Problems

1. Consider a group of 10 individuals with independent and identically distributed risks. They intend to experiment with different decentralized insurance schemes. Their common loss distribution is normal with mean 100 and variance 20.

   a. Suppose that the group self-organizes a mutual aid plan with no transaction cost. Whoever suffers a loss shall be fully compensated. The cost of compensation is carried equally by all members of the group. Calculate the mean and the variance of each member's cost under the mutual aid plan;
   b. Suppose that the group considers a P2P insurance scheme instead. Each member is willing to make a deposit of 110 to form a funding pool. They also find an insurer who agrees to pay any loss beyond the size of the funding pool. There is no profit margin for the insurer. Calculate the total premium for the excess-of-loss coverage by the insurer. Determine the mean and the variance of each member's cost under the P2P insurance scheme;
   c. Suppose that the group also works together under a simplified mudarabah takaful plan. Each member contributes 110 to a takaful fund at the beginning of each period. A takaful operator offers a qard hasan loan whenever there is a deficit in the takaful fund. The plan does not allow any reserve. In other words, any remaining balance shall be split between takaful members and the takaful operator. Determine the mudarabah rate at which the takaful operator breaks even on average. Determine the mean and the variance of each member's cost under the mudarabah takaful plan.
   d. Which decentralized scheme is the best if all members desire the smallest variance for their participation costs?

2. Consider the same group as in the previous question. The group decides to move from a mudarabah model to a wakalah model with consent from the takaful operator. What is the equivalent wakalah fee rate for the operator? Does the group reduce its variance by the switch?

Chapter 6
Aggregate Risk Pooling

The innovations of decentralized insurance can be found in different steps in an insurance process, from pricing, underwriting, to claims. The focus of this book is on new business models for risk sharing among various stakeholders. Such designs are based on fundamental actuarial concepts that go beyond the scope of traditional insurance. In this chapter, we consider a type of risk sharing scheme with a common feature that risks from all participants are aggregated and redistributed among participants. They frequently result from economic analysis of Pareto optimal preference measures for all participants. Despite the fact that there are many other types of risk sharing studied in the literature, this chapter offers three classes of risk sharing: utility-based, risk-measure-based, and conditional mean risk sharing.

6.1 Non-olet

While Chap. 5 offers various examples of existing practices on decentralized insurance, there is also a vast amount of academic literature on risk sharing. We shall explore their connections to the above-mentioned practices in Chap. 8. In this chapter, we discuss a range of classic risk sharing strategies that lead to the construction of the so-called "non-olet" risk pools. All members agree to pool together their risk-bearing capacity and share losses with each other according to some rule agreed upon prior to the observation of any loss. Although not imposed as a condition, classical risk sharing mechanisms turn out to have individual losses merged first before distributing the sum to individual members. The distribution of aggregate loss to participants does not distinguish the origin of losses.

The term "non-olet risk sharing" was coined by Borch (1960) and connected to a Roman anecdote on "pecunia non-olet". The anecdote goes back to the period of Emperor Vespasian (9 AD to 79) in the Roman Empire. The emperor instituted a

R. Feng, *Decentralized Insurance*, Springer Actuarial,
https://doi.org/10.1007/978-3-031-29559-1_6

(a) Public urinal in ancient Rome

(b) Vespasian gold coin

Fig. 6.1 Pecunia non-olet

urine tax on the distribution of urine. When the urine from Rome's public urinals (see Fig. 6.1a) was collected and sold for chemical processing, the tax was levied against the buyers of urine. His heir, Titan, found the notion of urine tax disgusting and challenged his father. Emperor Vespasian allegedly took out a gold coin and asked his son if the coin smelled (see Fig. 6.1b). When Titan answered no, Vespasian replied with the famous phrase "pecunia non-olet", which means money does not stink. It is used to suggest that the origin of the revenue does not taint its value. The term "non-olet" risk sharing is used by Borch to describe the feature of this type of risk sharing that the source of risk is irrelevant for the allocation of pooled risks.

Risk sharing is usually discussed in a setting of a heterogeneous pool. Suppose that there are n agents (individuals or entities) with different risks, denoted by $X_1, \cdots, X_n$. Risks are typically interpreted as non-negative loss random variables. Use the ith agent as a representative agent. The agent passes on the risk X_i to the pool and in exchange takes on a shared risk Y_i. There is no withdrawal from or deposit into the pool. Hence, the aggregate risk should be distributed to all agents, i.e.

$$\sum_{i=1}^{n} X_i = \sum_{i=1}^{n} Y_i, \qquad Y_i = h_i(S), \tag{6.1}$$

for some deterministic functions $h_1, \ldots, h_n$. This is often referred to as the loss conservation rule or the market clearing condition. Here the aggregate risk $S = \sum_{i=1}^{n} X_i$ is distributed in a single-blind way such that the amount of exchanged risk, X_i, carried by ith agent depends only on his or her own risk profile represented by h_i. In other words, all losses are first merged, and then divided up by h_i's. We shall refer to this type of risk sharing scheme a non-olet risk pool, or an aggregate risk pool, or a *risk pool* for short.

In Chap. 3, we discuss the optimal risk transfer from one party to another party in the context of traditional insurance, as all insurance contracts or reinsurance treaties are bilateral. In contrast, a decentralized insurance is always a multilateral agreement among a network of participants. Such an agreement is only sustainable if all participants reach consensus on how to share risks with each other. Therefore, we shall apply the Pareto optimality discussed in Chap. 3 to the settings of decentralized insurance.

The determination of risk pooling schemes is often based on some preference order. If Y is preferred over (or at least as preferred as) X, then we denote $Y \succ X$ ($Y \succcurlyeq X$). The preference can be measured by relative utility. Most risk pools in the literature are based on Pareto optimality, in which all agents try to maximize their preferences. A risk pool $(Y_1^*, \cdots, Y_n^*)$ is Pareto optimal if there is no other risk pool $(Y_1, \cdots, Y_n)$ that one agent is strictly better off and all other agents are no worse off, i.e.

- $\sum_{i=1}^n X_i = \sum_{i=1}^n Y_i$;
- $Y_i^* \succcurlyeq Y_i$ for all $i = 1, \cdots, n$;
- $Y_i^* \succ Y_i$ for at least one $i = 1, \cdots, n$.

Pareto optimal risk pools may not be unique or be worse than the original risks. Therefore, we often work with the Pareto optimal risk pool $(Y_1, \cdots, Y_n)$ that satisfies two additional properties.

- *(Individual rationality)* $Y_i \succcurlyeq X_i$ for all $i = 1, \cdots, n$;
- *(Actuarial fairness)* $\mathcal{P}[Y_i] = \mathcal{P}[X_i]$ for all $i = 1, \cdots, n$.

The individual rationality implies that every agent is better off participating in the risk pool than self-insuring. The actuarial fairness is needed to ensure that all agents are treated fairly under some premium principle $\mathcal{P}$. We often use the expected value principle as an example, i.e. $\mathcal{P}[X] = \mathbb{E}[X]$. However, other premium principles may also be used.

In the rest of this chapter, we shall consider three types of risk sharing rules using different preference orders.

- *Utility-Based Risk Sharing*
  The preference order $Y \succ X$ ($Y \succcurlyeq X$) is established by the expected disutility of taking on risk Y being less than (or no greater than) that of risk X, i.e. $\mathbb{E}[v(Y)] < \mathbb{E}[v(X)]$ ($\mathbb{E}[v(Y)] \leq \mathbb{E}[v(X)]$) for some disutility function v.

- *Risk-Measure-Based Risk Sharing*
  Risk measures are natural tools for gauging the effects of risk sharing rules. Assume that ρ is a risk measure with some desirable properties, such as value-at-risk and tail-value-at-risk. The preference order $Y \succ X$ ($Y \succcurlyeq X$) can be determined by the risk measure of the risk Y being less than (or no greater than) that of the risk X, i.e. $\rho[Y] < \rho[X]$ ($\rho[Y] \leq \rho[X]$).

- *Conditional Mean Risk Sharing*
  As the name suggests, the conditional mean risk sharing is defined by conditional means. Although not strictly equivalent, this type of risk sharing is related to the convex order as a preference order. In contrast to the utility-based risk sharing, the convex order requires the inequalities to hold for all disutility functions, not just one particular disutility function.

6.2 Utility-Based Risk Sharing

This section starts with the topic of risk sharing plan without any constraint. In such cases, there isn't any prescribed restriction on the possible outcome of risk sharing. Unrestricted risk sharing plan, however, could result in unlimited liabilities for participants which are undesirable for practical applications. The second part of the section is dedicated to the topic of Pareto optimal risk sharing with constraints. Detailed discussions of utility-based risk sharing can also be found in Borch (1968), Gerber and Pafum (1998), etc.

6.2.1 *Unconstrained Cases*

Consider n economic agents (individual, corporate, country, etc.) to share risks with each other. Let X_i be the risk of the ith agent prior to risk sharing for $i = 1, 2, \cdots, n$. Therefore, the risk vector prior to risk sharing is $(X_1, \ldots, X_n)$. The goal is to find a risk pool $(Y_1, \cdots, Y_n)$ so that

$$\sum_{i=1}^{n} X_i = \sum_{i=1}^{n} Y_i,$$

such that each agent is at least no worse off after risk sharing by the standards of expected utility. Assume that the utility function of the ith agent is denoted by u_i and the initial wealth of the agent is given by w_i. Therefore, the agent's expected utility prior to and post risk sharing are given by $\mathbb{E}[u_i(w_i - X_i)]$ and $\mathbb{E}[u_i(w_i - Y_i)]$, respectively.

Definition 6.1 (*Pareto Optimal Risk Exchange*) A risk pool $(Y_1, \cdots, Y_n)$ is said to be Pareto optimal if it is impossible to improve the situation of one agent without worsening that of at least one other agent. In other words, there is no other risk pool $(Z_1, Z_2, \cdots, Z_n)$ for which

$$\mathbb{E}[u_i(w_i - Y_i)] \leq \mathbb{E}[u_i(w_i - Z_i)], \qquad i = 1, \cdots, n,$$

where at least one of the inequalities is strict.

As discussed in Sect. 3.4, we usually look for Pareto optimal solutions by choosing some positive constants $k_1, k_2, \cdots, k_n$ and trying to maximize the objective function

$$\sum_{i=1}^{n} k_i \mathbb{E}[u_i(w_i - Y_i)],$$

for risk pool $(Y_1, \cdots, Y_n)$ such that

$$\sum_{i=1}^{n} X_i = \sum_{i=1}^{n} Y_i. \tag{6.2}$$

For brevity, we shall use the disutility function $v_i(Y_i) = -u_i(w_i - Y_i)$ instead of the utility function. Therefore, the same optimization can be rewritten as the minimization of the objective function

$$\sum_{i=1}^{n} k_i \mathbb{E}[v_i(Y_i)].$$

Theorem 6.1 *A risk pool $(Y_1, \cdots, Y_n)$ is Pareto optimal if and only if the random variables $k_i v_i'(Y_i)$ are the same for $i = 1, \cdots n$.*

Recall that the marginal rate of substitution is the rate at which a unit of one good can be traded for some amount of another good given the same utility. Let us think of the underlying risks as some commodities to be traded in this group of economic agents. We can define the overall utility of the entire group as $V(y_1, y_2, \cdots, y_n) = \sum_{i=1}^{n} k_i v_i(y_i)$ for any given scenario of loss $(y_1, \cdots, y_n)$. Then the marginal rate of substitution between any pair of risks is given by

$$\frac{V_{y_i}(y_1, \cdots, y_n)}{V_{y_j}(y_1, \cdots, y_n)} = \frac{k_i v_i'(y_i)}{k_j v_j'(y_j)}.$$

Theorem 6.1 suggests that the marginal rate of substitution should be the same for all pairs of risks under all scenarios. This result can be viewed as a stochastic extension of the equality (3.1) for Pareto optimality in the deterministic case.

It is also worthwhile pointing out that Pareto optimal risk sharing does not depend on the probability distribution of underlying risks, as the condition for the solution in Theorem 6.1 only refers to the functions $k_i v_i'$ for $i = 1, \cdots, n$. Therefore, the Pareto optimal risk sharing rule is the same for all cases of risk profiles as long as the participants' utility functions are known and fixed.

Example 6.1 (*Pareto Optimal Risk Exchange—Exponential Utility*) Suppose that in a risk sharing scheme there are n economic agents, each of whom uses an exponential utility function. Say, for $i = 1, \cdots, n$,

$$u_i(x) = \frac{1}{\alpha_i}(1 - e^{-\alpha_i x}),$$

where α_i is the constant risk aversion of the ith agent.

Assume that the ith agent's initial endowment is given by w_i. Every agent aims to maximize her own expected utility after the risk sharing. According to Theorem 6.1, in order to obtain the Pareto optimal risk exchange, we set, for all $i = 1, \cdots, n$,

$$k_i u_i'(w_i - Y_i) = k_i e^{-\alpha_i (w_i - Y_i)} = \Lambda,$$

which implies that

$$Y_i = \frac{\ln \Lambda}{\alpha_i} - \frac{\ln k_i}{\alpha_i} + w_i.$$

Observe that k_i is an arbitrary positive number. Each set of k_i's corresponds to a different Pareto optimal solution. We can write for some constant c_i

$$Y_i = \frac{\ln \Lambda}{\alpha_i} + c_i,$$

where the constants c_i's are also arbitrary pertaining to k_i's. Summing over all i's gives

$$S = \sum_{i=1}^{n} X_i = \sum_{i=1}^{n} Y_i = \frac{1}{\alpha} \ln \Lambda + \sum_{i=1}^{n} c_i,$$

where the constant α can be determined by

$$\frac{1}{\alpha} = \sum_{i=1}^{n} \frac{1}{\alpha_i}.$$

It follows immediately that

$$\ln \Lambda = \alpha S - \alpha \sum_{i=1}^{n} c_i.$$

Then, for some constants d_i's which can be determined later, we obtain

$$Y_i = \frac{\alpha}{\alpha_i} S + d_i.$$

Since $\sum_{i=1}^{n} Y_i = S$, then $\sum_{i=1}^{n} d_i = 0$. Again it shows that Pareto optimal solutions are not unique. To determine a unique Pareto optimal risk exchange, we impose the actuarial fairness condition

$$\mathbb{E}[X_i] = \mathbb{E}[Y_i].$$

Denote $\mathbb{E}[X_i] = \mu_i$. Then it is clear that

$$d_i = \mu_i - \frac{\alpha}{\alpha_i} \sum_{i=1}^{n} \mu_i .$$

Therefore, the fair Pareto optimal risk exchange is given by

$$Y_i = \mu_i + \frac{\alpha}{\alpha_i} (S - \mathbb{E}[S]) .$$

Observe that the optimal risk sharing is in fact a proportional risk transfer which was previously discussed in a bilateral setting in Sect. 3.3.4. Proportional risk transfer is in fact very common in various risk sharing plans to be discussed in the context of decentralized insurance.

Example 6.2 (*Pareto Optimal Risk Exchange—Power Utility*) Consider a risk sharing among a group of n economic agents. Suppose that each agent's satisfaction is measured by a power utility function. Say, for the ith agent, the utility function is given by

$$u_i(x) = \frac{x^{1-c} - 1}{1 - c}, \tag{6.3}$$

where $0 < c < 1$. To determine a Pareto optimal risk exchange, we set according to Theorem 6.1 that

$$k_i (w_i - Y_i)^{-c} = \Lambda,$$

which implies that, for some constant β_i,

$$Y_i = w_i - \beta_i \Lambda^{-1/c} .$$

Therefore, the aggregate loss S is given by

$$S = \sum_{i=1}^{n} w_i - \sum_{i=1}^{n} \beta_i \Lambda^{-1/c} .$$

It follows that

$$\Lambda^{-1/c} = \frac{1}{\sum_{i=1}^{n} \beta_i} \left(\sum_{i=1}^{n} w_i - S \right) .$$

Therefore, for some constant p_i, we have

$$Y_i = p_i S + \left(w_i - p_i \sum_{j=1}^{n} w_j \right) .$$

It is clear that $\sum_{i=1}^{n} p_i = 1$. Each set of p_i's corresponds to a different Pareto optimal solution. Note that the optimal risk sharing is a proportional risk transfer.

We find a unique risk sharing rule by considering the actuarial fairness condition. Then the constants p_i's can be uniquely determined by

$$p_i = \frac{w_i - \mu_i}{\sum_{j=1}^{n}(w_j - \mu_j)},$$

provided that the denominator is not zero. If we think of w_i as the capital investment of the ith agent to take on the underlying risk, then the quantity $w_i - \mu_i$ is interpreted as the "free capital" for the ith agent. The mean loss μ_i is in essence the agent's liability reserve. Any capital beyond the reserve is considered "free capital". It is interesting to see that the allocation of aggregate risk among agents is done in proportion to the agents' free capitals.

Example 6.3 (*Non-linear Pareto optimal risk sharing*) While risk sharing rules in both of the previous examples are linear in the aggregate risk S, that is not the case in general. Here we show an example of non-linear aggregate risk pooling rules. Consider two economic agents whose utility functions are given by $u_1(x) = \ln(x)$, $u_2(x) = 1 - 1/x$ for $x > 0$, which correspond to power utility functions in the form of (6.3) with $c = 1$ and $c = 2$. Let w_1 and w_2 be the initial wealth of the two agents, respectively. Suppose that the initial endowment of the two agents, $w_1 + w_2$, is sufficiently large to cover even worst scenarios, i.e. $S = X_1 + X_2 < w_1 + w_2$. Then, for some arbitrary positive constants $k_1 > 0$ and $k_2 > 0$, Pareto optimal risk sharing rules for these two agents are determined by

$$\frac{k_1}{w_1 - Y_1} = \frac{k_2}{(w_2 - Y_2)^2}.$$

Define $k = k_2/k_1 > 0$. Observe that solutions exist only when $w_1 > Y_1$. In view of the loss conservation $S = Y_1 + Y_2$, the equation can be written as

$$Y_2^2 - Y_2(2w_2 + k) + w_2^2 - w_1 k + Sk = 0.$$

Observe that $k^2 - 4Sk + 4w_1 k + 4kw_2 > 0$ by the assumption that $S < w_1 + w_2$. By solving the quadratic equation, we obtain the solution to Pareto optimal risk sharing

$$Y_1 = \frac{1}{2}(-k + 2S - 2w_2 + \sqrt{k^2 - 4Sk + 4w_1 k + 4kw_2});$$
$$Y_2 = \frac{1}{2}(k + 2w_2 - \sqrt{k^2 - 4Sk + 4w_1 k + 4kw_2}).$$

This solution also satisfies the condition that $w_1 > Y_1$ because

$$w_1 > \frac{1}{2}(-k + 2S - 2w_2 + \sqrt{k^2 - 4Sk + 4w_1k + 4kw_2}),$$

which is equivalent to

$$4(S - (w_1 + w_2))^2 > 0.$$

It should be pointed out that the Pareto optimal risk sharing exists for these two agents only if $S < w_1 + w_2$.

Note that the solution to Y_1 and Y_2 contains the arbitrary constant $k = k_2/k_1$, each value corresponding to a Pareto optimal risk sharing rule. We can determine a unique rule by the actuarial fairness condition, i.e.

$$\mathbb{E}[X_i] = \mathbb{E}[Y_i], \quad \text{for } i = 1, 2.$$

Let $\mathbb{E}[X_i] = \mu_i$ and $\mathbb{E}[S] = \mu_S$. The condition $\mathbb{E}[Y_2] = \mu_2$ implies that

$$(2\mu_2 - k - 2w_2)^2 = \left(\sqrt{k^2 - 4\mu_S k + 4w_1k + 4kw_2}\right)^2,$$

which determines a unique solution

$$k = \frac{(w_2 - \mu_2)^2}{w_1 - \mu_1}.$$

Note that in this case the optimal risk sharing is considered as a non-proportional risk transfer.

6.2.2 Constrained Cases

It is often the case that economic agents have limited risk-bearing capacity. For example, an agent may specify upon entrance into a risk sharing scheme that she can only accept losses up to a certain amount. Then it is necessary to consider the Pareto risk sharing with constraints. The solution given in this section is based on Bühlmann and Jewell (1979). There are some assumptions on this model. All companies have the same information about the statistical nature of the risk. Individual companies do not change their attitude according to the result.

The constrained case is a generalization of Theorem 6.1. Suppose that the economic agents share all their losses. Denote by X_i taken on by the ith agent prior to the risk sharing and Y_i the risk after the risk sharing. Note that all the loss must be covered, i.e. $\sum_i Y_i = \sum_i X_i = S$. It is often the case that we impose the following constraints on the risk sharing.

- *(Principle of Indemnity)* No one should profit from others' losses, i.e. $Y_i \geq 0$ for all $i = 1, \cdots, n$.
- *(Limited liability)* Each agent is held accountable up to some risk-bearing capacity, i.e. $Y_i \leq B_i$, where $B_i \geq 0$.

Recall that we shall use the disutility function for simplicity that $v_i(x) = -u_i(w_i - x)$. Note that, if each agent is risk averse, then $v_i' > 0$ and $v_i'' > 0$. The constrained risk sharing scheme can be determined as follows.

Theorem 6.2 *A risk pool $(Y_1, \cdots, Y_n)$ is Pareto optimal subject to the above-stated constraints if and only if there exists a vector of positive constants $(k_i)_{i=1}^n$ and positive random variable Λ for almost all outcomes such that*

$$\begin{aligned} k_i v_i'(Y_i) &= \Lambda \quad \text{if } 0 < Y_i < B_i \\ k_i v_i'(Y_i) &\geq \Lambda \quad \text{if } 0 = Y_i \\ k_i v_i'(Y_i) &\leq \Lambda \quad \text{if } Y_i = B_i. \end{aligned}$$

The proof is given in Bühlmann and Jewell (1978). Each set $(k_i)_{i=1}^n$ corresponds to a Pareto optimal solution in Theorem 6.2. Let us demonstrate how to use Theorem 6.2 to find a Pareto optimal solution.

For simplicity, assume that $B_i = \infty$. For effective upper bound, it can be handled similarly; see Example 6.7. It is often the case that a constrained Pareto optimal risk sharing scheme consists of layered, non-linear functions. Note that Λ is an intermediate random variable pertaining to losses Y_i's. We shall first set a fixed value $\Lambda = \lambda$, which works as a parameter to determine the allocation for each layer.

Example 6.4 Consider a three-agent risk sharing arrangement where $k_i v_i'(y)$ are known in advance and given by

$$\begin{aligned} k_1 v_1'(y) &= e^y, \\ k_2 v_2'(y) &= 3e^{y/3}, \\ k_3 v_3'(y) &= 5e^{y/5}. \end{aligned}$$

We shall visualize the procedure above in Fig. 6.2. By solving the equations, $k_i v_i'(y_i) = \lambda$ for $i = 1, 2, 3$, we obtain the inverse functions

$$\begin{aligned} y_1(\lambda) &= \ln \lambda, && 1 < \lambda; \\ y_2(\lambda) &= 3\ln(\lambda/3), && 3 < \lambda; \\ y_3(\lambda) &= 5\ln(\lambda/5), && 5 < \lambda. \end{aligned}$$

It follows from $s(\lambda) = \sum_{i=1}^3 y_i(\lambda)$ that

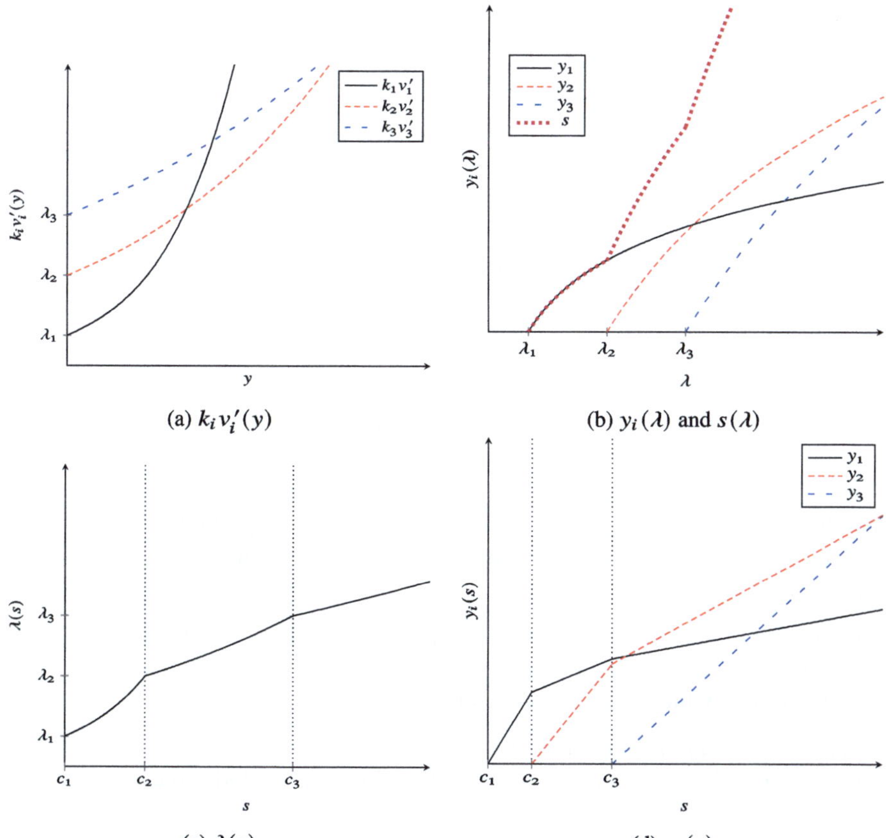

(a) $k_i v_i'(y)$ (b) $y_i(\lambda)$ and $s(\lambda)$

(c) $\lambda(s)$ (d) $y_i(s)$

Fig. 6.2 Intermediate variables in constrained risk sharing

$$s(\lambda) = \begin{cases} \ln \lambda, & 1 < \lambda < 3; \\ \ln (\lambda^4/3^3), & 3 < \lambda < 5; \\ \ln (\lambda^9/(3^3 5^5)), & 5 < \lambda. \end{cases}$$

Solving for the inverse function $\lambda(s)$ gives

$$\lambda(s) = \begin{cases} e^s, & 0 < s < \ln 3; \\ 3^{3/4} e^{s/4}, & \ln 3 < s < \ln(5^4/3^3); \\ 3^{1/3} 5^{5/9} e^{s/9}, & \ln(5^4/3^3) < s. \end{cases}$$

Inserting $\lambda(s)$ into $y_i(\lambda)$ for $i = 1, 2, 3$ yields the solution

$$y_1(s) = \begin{cases} s, & 0 < s < \ln 3; \\ \ln(3^{3/4}e^{s/4}), & \ln 3 < s < \ln(5^4/3^3); \\ \ln(3^{1/3}5^{5/9}e^{s/9}), & \ln(5^4/3^3) < s. \end{cases}$$

$$y_2(s) = \begin{cases} 0, & 0 < s < \ln 3; \\ 3\ln(3^{-1/4}e^{s/4}), & \ln 3 < s < \ln(5^4/3^3); \\ 3\ln(3^{-2/3}5^{5/9}e^{s/9}), & \ln(5^4/3^3) < s. \end{cases}$$

$$y_3(s) = \begin{cases} 0, & 0 < s < \ln 3; \\ 0, & \ln 3 < s < \ln(5^4/3^3); \\ 5\ln(3^{1/3}5^{-4/9}e^{s/9}), & \ln(5^4/3^3) < s. \end{cases}$$

Therefore, we split the aggregate risk S among the three agents by $(Y_1(S), Y_2(S), Y_3(S))$. Note that in all cases the agents' losses post risk sharing are always positive.

Following the previous example, we can formulate the solutions in a few steps. Define the inverse function of v_i' as γ_i, i.e. $(v_i')^{-1}(x) = \gamma_i(x)$. The risk sharing rule could be found by the following procedure.

1. *Set interval endpoints in the range of* λ.
   Given multipliers $\{k_i\}$ and the disutility functions $\{v_i\}$, re-index the agents so that $\lambda_1 \leq \lambda_2 \leq \ldots \leq \lambda_n$, where $\lambda_i = k_i v_i'(0)$. Take $\lambda_{n+1} = \infty$.

2. *Find* $y_i(\lambda)$ *as a piecewise-defined function.*
   For each instance of $\Lambda = \lambda$ such that $\lambda_j < \lambda \leq \lambda_{j+1}$, the agents with indices share the losses,

$$y_i(\lambda) = \begin{cases} \gamma_i\left(\frac{\lambda}{k_i}\right), & i = 1, 2, \cdots, j, \\ 0, & i = j+1, \cdots, n. \end{cases}$$

3. *Find* $s(\lambda)$ *and* $\lambda(s)$ *as piecewise-defined functions.*
   For $\lambda_j < \lambda \leq \lambda_{j+1}$, the total loss in this instance is given by

$$s(\lambda) = \sum_{i=1}^{n} y_i(\lambda) = \sum_{i=1}^{j} \gamma_i\left(\frac{\lambda}{k_i}\right).$$

   Since γ_i is an increasing function for each $i = 1, \cdots, n$, one can always find a unique inverse function $\lambda(s)$.

4. *Identify interval endpoints in the range of s.*
Define the ordered layering constants $c_1 = 0 \leq c_2 \leq c_3 \leq c_n$ with

$$c_i = s(\lambda_i) = \sum_{j=1}^{i-1} \gamma_j \left(\frac{\lambda_i}{k_j} \right),$$

where $i = 2, ..., n$. Note that $\lambda(s)$ is piecewise defined between any pair of these endpoints.

5. *Find $y_i(s)$ as a breakdown of s.*
With a slight abuse of notation, $y_i(s)$ is used to represent y_i as a function of s. For all $s < \sum_{i=1}^{n} y_i(\lambda_{j+1})$, it follows that for each $c_j \leq s < c_{j+1}$ and $j = 1, \cdots, n$,

$$y_i(s) = y_i(\lambda(s)),$$

where $y_i(\lambda)$ is defined in step 2 and $\lambda(s)$ is found in step 4.

Example 6.5 (*Positive Risk Sharing under Exponential Utility*) Let us revisit the example of exponential utility to obtain an explicit general solution in the procedure outlined above. Consider the ith agent's disutility function to be

$$v_i(x) = \alpha_i e^{x/\alpha_i},$$

for all $i = 1, \cdots, n$. In such a case, the endpoints in the range of λ are given by $\lambda_i = k_i v_i'(0) = k_i$ for $i = 1, \cdots, n$.. We first consider $y_1(\lambda)$ and $y_2(\lambda)$ as examples.

For $\lambda_1 \leq \lambda \leq \lambda_2$, we obtain

$$y_1(\lambda) = \alpha_1 \ln(\lambda/k_1), \qquad y_2(\lambda) = 0.$$

Therefore, for $\lambda_1 = k_1 < \lambda \leq \lambda_2 = k_2$,

$$s(\lambda) = y_1(\lambda).$$

Set the endpoints in the range of s by $c_1 = 0$ and $c_2 = s(\lambda_2) = \alpha_1 \ln(k_2/k_1)$. The inverse function is given for $c_1 < s \leq c_2$ by

$$\lambda(s) = e^{(s/\alpha_1)+\ln k_1}.$$

Thus, for $c_1 < s \leq c_2$, the breakdown of the aggregate risk can be given by

$$y_1(s) = y_1(\lambda(s)) = s, \qquad y_2 = 0.$$

For $\lambda_2 \leq \lambda \leq \lambda_3$, we obtain

$$y_1(\lambda) = \alpha_1 \ln(\lambda/k_1),\ y_2(\lambda) = \alpha_2 \ln(\lambda/k_2)$$

which leads to

$$s(\lambda) = \sum_{i\le 2} y_i(\lambda) = \alpha_1 \ln(\lambda/k_1) + \alpha_2 \ln(\lambda/k_2).$$

It shows that s lies between $c_2 = \alpha_1 \ln(k_2/k_1)$ and $c_3 = s(\lambda_3) = \alpha_1 \ln(k_3/k_1) + \alpha_2 \ln(k_3/k_2)$. Hence its inverse function is given for $c_2 < s \le c_3$ by

$$\lambda = e^{(s/(\alpha_1+\alpha_2))+(\alpha_1 \ln k_1+\alpha_2 \ln k_2)/(\alpha_1+\alpha_2)}.$$

Hence, for $c_2 < s \le c_3$, we obtain another breakdown of s

$$\begin{aligned}
y_1(s) &= \frac{\alpha_1}{\alpha_1+\alpha_2}s + \alpha_1(\alpha_1 \ln k_1 + \alpha_2 \ln k_2)/(\alpha_1+\alpha_2) - \alpha_1 \ln k_1 \\
&= \frac{\alpha_1}{\alpha_1+\alpha_2}s + \frac{\alpha_1\alpha_2}{\alpha_1+\alpha_2}(\ln k_2 - \ln k_1), \\
y_2(s) &= \frac{\alpha_2}{\alpha_1+\alpha_2}s + \alpha_2(\alpha_1 \ln k_1 + \alpha_2 \ln k_2)/(\alpha_1+\alpha_2) - \alpha_2 \ln k_2 \\
&= \frac{\alpha_2}{\alpha_1+\alpha_2}s - \frac{\alpha_1\alpha_2}{\alpha_1+\alpha_2}(\ln k_2 - \ln k_1).
\end{aligned}$$

To provide a concise but unified form, we can rewrite the solutions of y_1 and y_2 for $c_1 < s \le c_2$ and $c_2 < s \le c_3$ in the following way.

$$\begin{aligned}
y_1 &= \frac{\alpha_1}{\alpha_1}(s-c_1)_+ + \alpha_1\left(\frac{1}{\alpha_1+\alpha_2} - \frac{1}{\alpha_1}\right)(s-c_2)_+; \\
y_2 &= 0 \times (s-c_1)_+ + \frac{\alpha_2}{\alpha_1+\alpha_2}(s-c_2)_+.
\end{aligned}$$

By continuing this process for y_i with $j = 3, \cdots, n$ and $c_j < s \le c_{j+1}$ for $j = 3, \cdots, n$, we can find the general formula

$$\begin{aligned}
y_i(s) =& 0 \times (s-c_1)_+ + \cdots + \frac{\alpha_i}{\sum_{j=1}^{i}\alpha_j}(s-c_i)_+ + \alpha_i\left(\frac{1}{\sum_{j=1}^{i+1}\alpha_j} - \frac{1}{\sum_{j=1}^{i}\alpha_j}\right)(s-c_{i+1})_+ \\
&+ \cdots + \left(\alpha_i\left(\frac{1}{\sum_{j=1}^{n}\alpha_j} - \frac{1}{\sum_{j=1}^{n-1}\alpha_j}\right)\right)(s-c_n)_+.
\end{aligned}$$

Therefore, written in the most concise form, the allocation $y_i(s)$ of the total loss s, for $i = 1, 2, \cdots, n$,

$$y_i(s) = \sum_{j=i}^{n} \beta_{ij}(s-c_j)_+,$$

with $c_1 = 0 \le c_2 \le \cdots \le c_n$, and

$$\beta_{ij} = \begin{cases} 0, & j < i; \\ \alpha_i / \sum_{k=1}^{i} \alpha_k, & j = i; \\ \alpha_i [(\sum_{k=1}^{j} \alpha_k)^{-1} - (\sum_{k=1}^{j-1} \alpha_k)^{-1}], & j > i, \end{cases}$$

independent of c_j's. In the exponential case, the economic agents quota-share in layers, where the quota for each agent is calculated by dividing their risk aversion rate with the sum of risk aversion rates of all participants in that layer.

Example 6.6 We revisit Example 6.4 to write the piecewise-defined expressions that can be written in a more concise form. Note, for example, if $\ln 3 < s < \ln(5^4/3^3)$,

$$y_1(s) = \ln(3^{3/4} e^{s/4}) = \frac{1}{4} s + \frac{3}{4} \ln(3) = s - \frac{3}{4}(s - \ln(3))_+.$$

Similarly, the solution to the three-agent arrangement can be rewritten as

$$y_1(s) = s - \frac{3}{4}\left(s - \ln(3)\right)_+ - \frac{5}{36}\left(s - \ln(5^4/3^3)\right)_+,$$
$$y_2(s) = \frac{3}{4}\left(s - \ln(3)\right)_+ - \frac{5}{12}\left(s - \ln(5^4/3^3)\right)_+,$$
$$y_3(s) = \frac{5}{9}\left(s - \ln(5^4/3^3)\right)_+.$$

In previous examples, the risk sharing rules are usually not unique and determined by arbitrary positive constants $\boldsymbol{k} = (k_1, k_2, \cdots, k_n)$, or equivalently by its layering constants $\boldsymbol{c} = (c_1, c_2, \cdots, c_n)$. Each set of $\boldsymbol{k}$ or equivalently $\boldsymbol{c}$ shows a particular Pareto optimal solution. In order to determine a unique Pareto optimal solution, we use the criterion for actuarial fairness. For all $i = 1, 2, \cdots, n$, we set

$$\mathcal{P}\left[y_i(S; \boldsymbol{c})\right] = \mathcal{P}[X_i] \quad \text{or} \quad \mathcal{P}\left[y_i(S; \boldsymbol{k})\right] = \mathcal{P}[X_i],$$

where we add $\boldsymbol{c}$ and $\boldsymbol{k}$ as a parameter to indicate the dependence of the solution on $\boldsymbol{c}$ and $\boldsymbol{k}$.

It is worthwhile pointing out that the premium principles may not always lead to explicit expressions and one may have to resort to numerical solutions. When performing a root search procedure, the following observations could be helpful. By the loss conservation condition, it holds that $\sum_i \mathcal{P}\left[y_i(S; \boldsymbol{k})\right] = \sum_i \mathcal{P}[X_i]$. Note that as k_j increases with all else the same, the jth agent enters in the redistribution of loss at a higher λ_j. Thus, the jth agent takes less loss than otherwise. By the loss conservation condition, the rest of the aggregate loss shall be redistributed to other participants. In other words, if k_j increases, then $\mathcal{P}\left[y_i(X; \boldsymbol{k})\right]$ is non-decreasing for $i \neq j$ and it is non-increasing for $i = j$.

Example 6.7 Consider a two-agent risk sharing problem where the first agent is only willing to accept risk up to an upper bound, $y_1 \le B_1$ for $B_1 > 0$, and the second agent takes the residual risk. Assume that

$$\begin{aligned} k_1 v_1'(y_1) &= k_1 e^{y_1}, \\ k_2 v_2'(y_2) &= k_2 e^{y_2/\alpha}, \end{aligned}$$

where $y_1, y_2 > 0$, and k_1, k_2 are constants such that $k_1 e^{B_1} > k_2 > k_1 > 0$. By Theorem 6.2, the Pareto optimal risk exchange can be determined by first solving the equations $k_i v_i'(y_i) = \lambda$ for $i = 1, 2$. We shall represent y_1, y_2 as functions of λ. Observe that

$$y_1(\lambda) = \begin{cases} 0 & 0 < \lambda < k_1 \\ \ln \lambda / k_1 & k_1 < \lambda < k_1 e^{B_1} \\ B_1, & k_1 e^{B_1} < \lambda \end{cases};$$

$$y_2(\lambda) = \begin{cases} 0 & 0 < \lambda < k_2 \\ \alpha \ln(\lambda/k_2), & k_2 < \lambda \end{cases}.$$

It follows from $s(\lambda) = y_1(\lambda) + y_2(\lambda)$ that

$$s(\lambda) = \begin{cases} \ln(\lambda/k_1), & k_1 < \lambda < k_2; \\ \ln(\lambda^{\alpha+1}/k_1 k_2^{\alpha}), & k_2 < \lambda < k_1 e^{B_1}; \\ B_1 + \alpha \ln(\lambda/k_2), & k_1 e^{B_1} < \lambda. \end{cases}$$

Since B_1 is big enough, we can let $B_1 = \ln(k_2/k_1) + b$ for some $b > 0$. Define $c = (\alpha + 1)b + \ln(k_2/k_1)$. Solving for the inverse function $\lambda(s)$ gives

$$\lambda(s) = \begin{cases} k_1 e^s, & 0 < s < \ln(k_2/k_1); \\ (k_1 k_2^{\alpha} e^s)^{1/(\alpha+1)}, & \ln(k_2/k_1) < s < c; \\ k_2 e^{(s-B_1)/\alpha}, & c < s. \end{cases}$$

Define $k = \ln(k_2/k_1)$. Inserting $\lambda(s)$ into $y_i(\lambda)$ for $i = 1, 2$ yields the solution

$$y_1(s) = \begin{cases} s, & 0 < s < k; \\ (1/(\alpha+1))s + (\alpha/(\alpha+1))k, & k < s < c; \\ B_1, & c < s. \end{cases}$$

$$y_2(s) = \begin{cases} 0, & 0 < s < k; \\ (\alpha/(\alpha+1))s - (\alpha/(\alpha+1))k, & k < s < c; \\ s - B_1, & c < s. \end{cases}$$

Suppose that two insurers have the risk profiles X_1 and X_2. Suppose that the total loss $S = X_1 + X_2$ follows the exponential distribution $f_S(s) = (1/\delta)e^{-(1/\delta)s}$. Note

$$\begin{aligned} \mathbb{E}\left[y_1(S;k)\right] &= \frac{1}{\delta}\int_0^k s e^{-(s/\delta)}\,\mathrm{d}s + \frac{1}{\delta}\int_k^c \left(\frac{1}{\alpha+1}s + \frac{\alpha}{\alpha+1}k\right)e^{-(s/\delta)}\,\mathrm{d}s \\ &\quad + \frac{1}{\delta}\int_c^\infty B_1 e^{-(s/\delta)}\,\mathrm{d}s \\ &= \delta - \frac{\alpha\delta}{\alpha+1}e^{-(k/\delta)} - \frac{\delta}{\alpha+1}e^{-(c/\delta)}; \\ \mathbb{E}\left[y_2(S;k)\right] &= \frac{1}{\delta}\int_k^c \left(\frac{\alpha}{\alpha+1}s - \frac{\alpha}{\alpha+1}k\right)e^{-(s/\delta)}\,\mathrm{d}s \\ &\quad + \frac{1}{\delta}\int_c^\infty (s - B_1)e^{-(s/\delta)}\,\mathrm{d}s. \\ &= \frac{\alpha\delta}{\alpha+1}e^{-(k/\delta)} + \frac{\delta}{\alpha+1}e^{-(c/\delta)}, \end{aligned}$$

where they sum up to δ as expected from $\mathbb{E}\left[S\right] = \delta = \mathbb{E}\left[y_1(S;k)\right] + \mathbb{E}\left[y_2(S;k)\right]$. By the fair premium principle, k should be determined to satisfy

$$\begin{aligned} \mathbb{E}\left[y_1(S;k)\right] &= \mathbb{E}[X_2] = \mu_1; \\ \mathbb{E}\left[y_2(S;k)\right] &= \mathbb{E}[X_2] = \mu_2. \end{aligned}$$

We can use either equation to fix k since $\delta = \mathbb{E}\left[y_1(S;k)\right] + \mathbb{E}\left[y_2(S;k)\right]$. It follows from the equation $\mathbb{E}[y_2(S;k)] = \mu_2$ that

$$\frac{\alpha\delta}{\alpha+1}e^{-(k/\delta)} + \frac{\delta}{\alpha+1}e^{-(c/\delta)} = \mu_2.$$

Solving the equation above for k leads to

$$k = \delta\left(\log(\alpha + e^{-(\alpha+1)b/\delta}) - \log\mu_2 - \log(\alpha+1) + \log\delta\right).$$

One could verify that k also satisfies the other equation $\mathbb{E}[y_1(S;k)] = \mu_1$.

6.3 Risk-Measure-Based Risk Sharing

While expected utility is a popular tool to analyze an economic agent's decisions in face of uncertainty, some argue that extreme events are the most important factors for decision-making from the perspective of risk management. Therefore, risk measures have also been used in the literature to determine the preference order of risk sharing strategies.

Consider the case again where the risks from n economic agents, $(X_1, \cdots, X_n)$, are pooled together and the agents seek to identify an optimal plan $(Y_1, \cdots, Y_n)$ to share the aggregate risk. In a one-period model, the risk sharing strategy should meet the loss conservation condition

$$\sum_{i=1}^{n} X_i = \sum_{i=1}^{n} Y_i.$$

In search for an optimal plan, each agent is known to prefer as low-risk measure as possible, say, $\mathrm{VaR}_{\alpha_i}[Y_i]$, for the ith agent for $i = 1, \cdots, n$. In other words, we are interested in devising a risk sharing plan that minimizes every agent's VaR risk measure in the Pareto setting.

Definition 6.2 (*Pareto Optimal Risk Pool*) A risk pool $(Y_1, \cdots, Y_n)$ is said to be Pareto optimal if it is impossible to improve the situation of one agent without worsening that of at least one other agent. In other words, there is no other risk pool $(Z_1, Z_2, \cdots, Z_n)$ for which

$$\mathrm{VaR}_{\alpha_i}[Z_i] \le \mathrm{VaR}_{\alpha_i}[Y_i], \qquad i = 1, \cdots, n,$$

where at least one of the inequalities is strict.

The discussion in this section is based on the work of Embrechts et al. (2018). It is shown in their work that the Pareto optimal risk pool is obtained if we can find $(Y_, \cdots, Y_n)$ that minimizes the objective function

$$\sum_{i=1}^{n} \mathrm{VaR}_{\alpha_i}[Y_i]. \tag{6.4}$$

Note, however, unlike the Pareto optimality discussed earlier, we cannot seek to minimize a weighted sum of VaRs. Take the case of two agents as an example. Since VaR is translation invariant, we can always make the weighted sum of VaRs smaller by shifting a constant amount of cash between two agents. Assume (Y_1, Y_2) is a given risk sharing rule. We can create another risk sharing rule by setting $Z_1 = Y_1 - m$ and $Z_2 = Y_2 + m$. Observe that

$$w_1 \mathrm{VaR}_{\alpha_1}[Z_1] + w_2 \mathrm{VaR}_{\alpha_2}[Z_2] = w_1 \mathrm{VaR}_{\alpha_1}[Y_1] + w_2 \mathrm{VaR}_{\alpha_2}[Y_2] + (w_2 - w_1)m.$$

We can choose m such as $(w_2 - w_1)m < 0$ and obtain

$$w_1 \mathrm{VaR}_{\alpha_1}[Z_1] + w_2 \mathrm{VaR}_{\alpha_2}[Z_2] \leq w_1 \mathrm{VaR}_{\alpha_1}[Y_1] + w_2 \mathrm{VaR}_{\alpha_2}[Y_2].$$

Therefore, it is clear that there is no risk sharing rule that minimizes the weighted sum of VaR risk measures unless the weights are equal as in (6.4).

As the focus of the optimization is on extreme events, we can consider the $100\hat{\alpha}_i\%$ worst scenarios where $\hat{\alpha}_i = 1 - \alpha_i$ for $i = 1, \cdots, n$. Assume that $\sum_{k=1}^{i} \hat{\alpha}_k \leq 1$ and $S = \sum_{i=1}^{n} X_i$ has a continuous distribution.

Theorem 6.3 *A Pareto optimal risk pool $(Y_1, \cdots, Y_n)$ with respect to VaR risk measure is given by*

$$Y_i = (S - m)I(\gamma_i < S \leq \gamma_{i-1}), \qquad i = 1, \cdots, n-1; \tag{6.5}$$

$$Y_n = (S - m)I(S \leq \gamma_{n-1}) + m, \tag{6.6}$$

where γ_i is determined by

$$\gamma_i = \mathrm{VaR}_{1-\sum_{k=1}^{i} \hat{\alpha}_k}[S],$$

and m is an arbitrary number such that $m \leq \gamma_n$.

The proof is given in Appendix 6.C. This result shows that the Pareto optimal risk pool is in essence an multi-party extension of the non-proportional risk transfer with franchise deductible in a bilateral insurance contract in Sect. 3.1.2. A visualization of this excess-of-loss treaty is given in Fig. 6.3. When $m = 0$, the split of the aggregate risk is straightforward. The nth participant takes on the aggregate risk when it is in the bottom range $(-\infty, \gamma_n)$. The $(n-1)$th participant accepts the aggregate risk in the range (γ_n, γ_{n-1}). The pattern continues until the first participant is left with the risk in the top range (γ_1, ∞). In other words, when the loss is very small, the loss is carried by the nth participant until it reaches his or her liability limit γ_n. As the size of loss increases, the risk is passed on to the next participant until his or her limit is reached, in the order from the $n-1$th participant to the 1st participant. Note that this is not an excess-of-loss policy, where only excessive amount is passed on to the next participant. There is only one participant taking on the entire amount of loss in every scenario. When $m \neq 0$, the nth participant takes on the aggregate risk when it is in the range $(-\infty, \gamma_n)$ or the amount m when it exceeds γ_n. In each of the next ranges, one participants takes the risk less the amount m. In other words, a fixed amount of loss m (or profit if $m < 0$) is carried by the nth participant. The rest is passed around among the participants depending on the size of the loss. When the aggregate risk is always positive, then one can always choose $m = 0$.

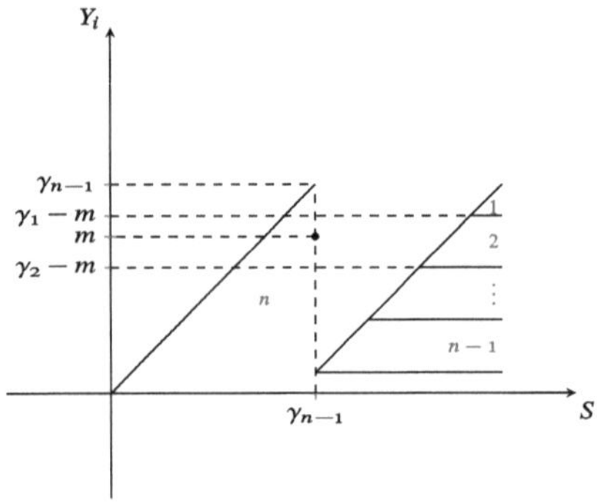

(a) when $m > 0$

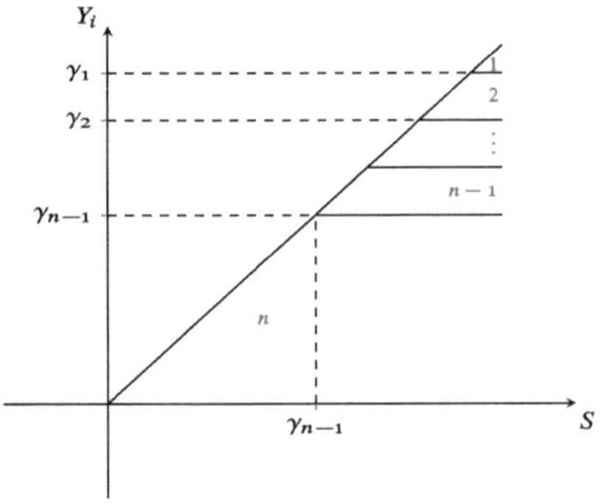

(b) when $m = 0$

Fig. 6.3 Quantile risk-measure-based risk sharing

As indicated above, Pareto optimal risk pool is not unique with respect to the quantile risk measure. We can identify a unique risk pool by the actuarial fairness requirement. Since VaRs are translation invariant, the sum of VaRs does not change if we allow side payments among all participants as long as the sum of side payments is zero. Hence, the fair Pareto optimal risk pool is determined by, for $i = 1, \cdots, n-1$,

$$Y_i = (S - m)I(\gamma_i < S \le \gamma_{i-1}) + \mathbb{E}[X_i] - \mathbb{E}[(S - m)I(\gamma_i < S \le \gamma_{i-1})], \quad (6.7)$$
$$Y_n = (S - m)I(S \le \gamma_{n-1}) + \mathbb{E}[X_n] - \mathbb{E}[(S - m)I(S \le \gamma_{n-1})]. \quad (6.8)$$

Here we provide an example of the quantile-based risk sharing of normally distributed risks.

Example 6.8 Suppose that there are n participants in a risk exchange, where the ith participant brings in the normally distributed risk X_i with the mean μ_i and the variance σ_i^2, and all risks are independent of each other. Assume that the ith participant uses the risk measure VaR_{α_i} for his/her decision-making. Note that the sum of the risk S follows the normal distribution with the mean $\sum_{i=1}^n \mu_i$ and the variance $\sum_{i=1}^n \sigma_i^2$. Recall that $\mathrm{VaR}_p(X) = \mu + \sigma\Phi^{-1}(p)$ for $0 < p < 1$ and X is normally distributed with mean μ and the variance σ^2. Then we must have

$$\gamma_i = \mathrm{VaR}_{1-\sum_{k=1}^i \hat{\alpha}_k}[S] = \sum_{i=1}^n \mu_i + \Phi^{-1}\left(1 - \sum_{k=1}^i \hat{\alpha}_k\right)\sqrt{\sum_{i=1}^n \sigma_i^2}.$$

First, consider the post-exchange risk of the nth agent. Observe that, for some arbitrary m such that $m \le \gamma_n$,

$$\mathbb{E}[(S-m)I(S \le \gamma_{n-1})] = \mathbb{P}(S \le \gamma_{n-1})\left(\mathbb{E}[S|I(S \le \gamma_{n-1})] - m\right)$$
$$= \left(1 - \sum_{k=1}^{n-1} \hat{\alpha}_k\right)\left(\sum_{i=1}^{n} \mu_i - \sqrt{\sum_{i=1}^{n} \sigma_i^2} \frac{\phi(\beta_n)}{\Phi(\beta_n)} - m\right),$$

where

$$\beta_n = \frac{\gamma_{n-1} - \sum_{i=1}^{n} \mu_i}{\sqrt{\sum_{i=1}^{n} \sigma_i^2}} = \Phi^{-1}\left(1 - \sum_{k=1}^{n-1} \hat{\alpha}_k\right).$$

Therefore, the post-exchange risk for the nth participant is provided by

$$Y_n = (S-m)I(S \le \gamma_{n-1}) + \mu_n - \left(1 - \sum_{k=n-1}^{i} \hat{\alpha}_k\right)\left(\sum_{i=1}^{n} \mu_i - \sqrt{\sum_{i=1}^{n} \sigma_i^2} \frac{\phi(\beta_n)}{\Phi(\beta_n)} - m\right).$$

Next, we consider the risk sharing for the ith participant for $i = 1, 2, \cdots, n-1$. Observe that

$$\mathbb{E}[(S-m)I(\gamma_i < S \le \gamma_{i-1})] = \mathbb{P}(\gamma_i < S \le \gamma_{i-1})\left(\mathbb{E}[S|I(\gamma_i < S \le \gamma_{i-1})] - m\right)$$
$$= \hat{\alpha}_i \left(\sum_{i=1}^{n} \mu_i - \sqrt{\sum_{i=1}^{n} \sigma_i^2} \frac{\phi(\beta_i) - \phi(\beta_{i+1})}{\Phi(\beta_i) - \Phi(\beta_{i+1})} - m\right).$$

Therefore, the post-exchange risk for the ith participant is given by

$$Y_i = (S-m)I(\gamma_i < S \le \gamma_{i-1}) + \mu_i - \hat{\alpha}_i \left(\sum_{i=1}^{n} \mu_i - \sqrt{\sum_{i=1}^{n} \sigma_i^2} \frac{\phi(\beta_i) - \phi(\beta_{i+1})}{\Phi(\beta_i) - \Phi(\beta_{i+1})} - m\right).$$

6.4 Conditional Mean Risk Sharing

One could also use the convex order as a preference order for decision-making. Recall that $Y \le_{cx} X$ if $\mathbb{E}[v(Y)] \le \mathbb{E}[v(X)]$ for all convex functions v. In this case, we say that the risk Y is preferred over the risk X, i.e. $Y \succeq X$, if a risk-averse individual's expected utility undertaking the risk Y is always higher than that undertaking the risk X, i.e. $\mathbb{E}[u(w-Y)] \ge \mathbb{E}[u(w-X)]$ for any arbitrary concave utility function u and any arbitrary initial wealth w. The function v can be viewed as a disutility function corresponding to the utility function u. Note that such a preference order is more strict than that established by the expected utility for a specific utility function in Sect. 6.2. Motivated by the convex order, Denuit and Dhaene (2012a) introduced another type of risk sharing rule, known as *conditional mean risk sharing*.

Definition 6.3 The conditional mean risk sharing rule is defined by, for all $i = 1, \cdots, n$,

$$Y_i = h_i(S) = \mathbb{E}[X_i|S].$$

By the nature of conditional expectation, the actuarial fairness is automatically satisfied for $\mathcal{P} = \mathbb{E}$, i.e. $\mathbb{E}[Y_i] = \mathbb{E}[X_i]$, for $i = 1, \ldots, n$. Observe that $\sum_{i=1}^{n} Y_i = \sum_{i=1}^{n} \mathbb{E}[X_i|S] = \mathbb{E}[S|S] = S = \sum_{i=1}^{n} X_i$, which indicates that it is a risk sharing rule.

Reconsider the situation where economic agents with their own risks $(X_1, \cdots, X_n)$ negotiate with each other for a risk sharing solution. Assume that all agents are risk averse. The following result shows that there always exists a mutually beneficial risk sharing mechanism with respect to the convex order. Regardless of their utility functions, the agents can also increase their respective expected utilities by engaging in a conditional mean risk sharing. Furthermore, if $\mathbb{E}[X_1|S], \cdots, \mathbb{E}[X_n|S]$ are comonotonic, then the conditional mean risk allocation is Pareto optimal. The proof of this result can be found in Appendix 6.B.

Theorem 6.4 *For all risks X_i's, the conditional mean risk allocation is mutually beneficial for all participants in convex order. That is,*

$$h_i(S) \leq_{cx} X_i \quad \text{for } i = 1, 2, \ldots, n.$$

If $h_i(S)$'s are comonotonic, then the conditional mean risk allocation is Pareto optimal.

The connection of comonotonic risk sharing with Pareto optimality in convex order is known in Carlier et al. (2012).

An interesting special case is that the underlying risks of all agents are exchangeable. In other words, all risks have the same marginal distribution but may have arbitrary dependence among each other. It is clear that $h_i(S) = \mathbb{E}[X_i|S] = \mathbb{E}[X_j|S] = h_j(S)$. Hence, it must be true that $h_i(S) = (1/n)S$. In such a case, all individuals share the total losses equally. Since the risk allocation $h_i(s) = (1/n)s$ is clearly non-decreasing, it follows from Theorem 6.4 that the conditional mean risk sharing is uniform and Pareto optimal for all exchangeable risks.

Example 6.9 (*Conditional Mean Risk Sharing for Normal Risks*) Suppose that the risks $(X_1, \cdots, X_n)$ are normally distributed. Note from Example 2.17 that

$$Y_i = \mathbb{E}[X_i|S] = \mathbb{E}[X_i] + \frac{\mathbb{C}[X_i, S]}{\mathbb{V}[S]}(S - \mathbb{E}[S]).$$

It implies that the conditional mean risk sharing is linear for normal risks.

Consider a numerical example for the two-agent case. The risks brought by the two agents are assumed to be independent of each other and $\mu_1 = 4, \sigma_1 = 2, \mu_2 = 6, \sigma_2 = 4$. It is clear that $\mathbb{C}[X_i, S] = \mathbb{C}[X_i, X_i] = \sigma_i^2$. In the case that the total loss happens to be $S = 15$, we can then allocate the total loss between the two agents in the following way.

$$h_1(15) = \mu_1 + \frac{\sigma_1^2}{\sigma_1^2 + \sigma_2^2}\Big(S - \mu_1 - \mu_2\Big) = 5;$$

$$h_2(15) = \mu_2 + \frac{\sigma_2^2}{\sigma_1^2 + \sigma_2^2}\Big(S - \mu_1 - \mu_2\Big) = 10.$$

Example 6.10 (*Conditional Mean Risk Sharing for Mutually Exclusive Risks*) The non-negative discrete risks $(X_1, ..., X_n)$ are mutually exclusive if

$$P[X_i > 0, X_j > 0] = 0 \quad \text{for all} \quad i \neq j.$$

In other words, if the risks from all economic agents are mutually exclusive, one can only observe positive loss from one agent at a time. For $s > 0$ such that $\mathbb{P}[S = s] > 0$, we have

$$h_i(s) = \mathbb{E}[X_i|S = s] = \sum_{j=1}^{n} \mathbb{E}[X_i|S = s, X_j > 0]\mathbb{P}[X_j > 0|S = s] = s\,\mathbb{P}[X_i > 0|S = s].$$

Therefore, the conditional mean risk sharing rule is given by

$$h_i(s) = \frac{\mathbb{P}[X_i > 0, S = s]}{\sum_j \mathbb{P}[X_j > 0, S = s]}\,s.$$

For simplicity, we further assume that $\mathbb{P}[X_i > 0] = p_i$ and $\mathbb{P}[X_i = 0] = 1 - p_i$ for all i. Then, $\mathbb{P}[X_i > 0, S = s] = p_i$. Thus,

$$h_i(S) = \frac{p_i}{\sum_i p_i} S, \quad S \in \{0, s\}.$$

It is clear that the conditional mean risk sharing is quota-share for mutually exclusive risks. Since each of h_i's is also a non-decreasing function, the $h_i(S)$'s are comonotonic. Hence, the risk sharing is Pareto optimal for all economic agents in the sense of convex ordering.

More examples of conditional mean risk sharing with explicit solutions can be found in Denuit (2019, 2020), Denuit et al. (2022b), etc.

6.5 Visualization in Cake-Cutting

The process of risk pooling and allocation has some analogy with that of cake making and cutting. Many ingredients, such as flour, sugar, eggs, and butter, go into the making of a cake, as much as all types of risks are mixed together in the setup of a risk pool. While an experienced baker can make a fancy cake, it requires a different skill to cut the cake to serve people well in a party. There are many intuitive ways of cutting a cake. But the problem of dividing a cake fairly for an arbitrary number of eaters is a complex scientific issue and some mathematicians have developed fairly sophisticated algorithms to tackle such a problem. Just googling "envy-free cake-cutting" should lead to many research works. The allocation of pooled risks to risk takers is arguably an even more complex problem, as the "cake" in question here is the aggregate risk, which is random in nature. The interpretation of "fairness" for such cake-cutting can be different by risk takers.

We can use the cake-cutting metaphor to help visualize several risk sharing strategies developed in this chapter. One can think of the size of each layer of a cake as a representation of a particular scenario of the aggregate loss. The cake is made up of infinite many layers of possible scenarios. In other words, we represent loss severity in the horizontal space and probability in the vertical direction.

Under the utility-based risk sharing framework, pro rata risk sharing is known to be Pareto optimal in several unconstrained cases with various utility functions, such as exponential, power, and logarithmic utilities. Such an approach is in essence to cut the cake in wedges as shown in Fig. 6.4. Regardless of the loss severity, each participant carries a fixed percentage of the aggregate loss.

In contrast, the Pareto optimal VaR-based risk sharing strategy is to partition the full probability set into mutually exclusive sets and to require each participant to

Fig. 6.4 Concept visualization of pro rata risk sharing

Fig. 6.5 Concept visualization of VaR-based optimal risk sharing

Fig. 6.6 Concept visualization of conditional mean risk sharing

take all of the loss (or adjusted by a fixed amount) in each given set of possibilities. Metaphorically, one can think of it as a method to cut the cake in layers of possibility, each of which is given to a participant, as shown in Fig. 6.5. Each layer represents that one party takes it all in a set of scenarios.

The visualization of conditional mean risk sharing is a little tricky. The ith participant receives the share $h_i(s)$ if the aggregate loss turns out to be $S = s$. In other words, the shared percentage for a given participant varies case by case. To illustrate this idea in a simple way, we can think of cutting the cake in different ways for various layers. Suppose that there are two parties to share the aggregate loss in the pool and that the function $h_1(s)$ is an increasing function of s for scenarios $S = s$ from the top to bottom. Each participant receives a stack of slices cut with some angle, as illustrated in Fig. 6.6. Then the first participant collects the left stack of layers and the rest goes to the second participant.

The discussion in this section is largely figurative and only used for illustrative purposes to help readers understand the abstract concept of risk sharing.

There are also a wide variety of other risk sharing rules derived from the optimization of risk measures, including quantile-based risk sharing in Embrechts et al. (2018), inf-convolution of convex risk measure in Barrieu and Karoui (2005), law-invariant monetary utility functions in Jouini et al. (2008), etc.

Bear in mind that, while we have introduced a variety of aggregate risk pooling rules in explicit forms, it is not easy to derive exact formulas for risk sharing rules under general assumptions. Another criticism of many aggregate risk pooling rules is that cash transfers occur even if nobody makes any claim, i.e. $h_i(0) \neq 0$. Such a property may not be viewed as desirable, as payments are forced upon participation. In such case, the amount $h_i(0)$ is sometimes referred to as a side payment. Some argue that the side payment can be considered a "premium" that one pays to participate in a risk sharing plan.

In the next chapter, we shall discuss quota-share rules, which by design have relatively simple forms and require no ex-ante side payment or "premium".

Appendix 6.A Pareto Optimal Risk Exchange Condition

Proof Suppose that $(Y_1, \cdots, Y_n)$ is Pareto optimal. Define, for a real number t, any pair $j, h = 1, \cdots, n$, and an arbitrary random variable V,

$$\begin{cases} \hat{Y}_i = Y_i, & i \neq j, h \\ \hat{Y}_j = Y_j + t\Delta, \\ \hat{Y}_h = Y_h - t\Delta. \end{cases}$$

Consider the function

$$f(t) = \sum_{i=1}^{n} k_i \mathbb{E}[v_i(Y_i)].$$

Since the function is assumed to be minimized at $t = 0$, then we must have $f'(0) = 0$, which implies that

$$k_j \mathbb{E}[\Delta v_j'(Y_j)] = k_h \mathbb{E}[\Delta v_h'(Y_h)].$$

Because the equality holders for any arbitrary Δ, we can conclude that

$$k_j v_j'(Y_j) = k_h v_h'(Y_h).$$

Note that the choices of j and h are also arbitrary. Therefore, the equality must hold for all $j, h = 1, \cdots, n$.

Suppose that $(Y_1, \cdots, Y_n)$ is a risk exchange for which $k_j v_j'(Y_j)$ is the same for all $j = 1, \cdots, n$. We can show that $(Y_1, \cdots, Y_n)$ is Pareto optimal. Note that a convex curve is always above its tangent lines, i.e. for all x, y

$$v(x) \geq v(y) + v'(y)(x - y).$$

Let $(X_1, \cdots, X_n)$ be any other risk exchange. Then for all i,

$$v_i(X_i) \geq v_i(Y_i) + v_i'(Y_i)(X_i - Y_i).$$

Therefore,

$$\begin{aligned} \sum_{i=1}^{n} k_i v_i(X_i) &\geq \sum_{i=1}^{n} k_i v_i(Y_i) + \sum_{i=1}^{n} k_i v_i'(Y_i)(X_i - Y_i) \\ &= \sum_{i=1}^{n} k_i v_i(Y_i) + k_1 v_1'(Y_1) \sum_{i=1}^{n} (X_i - Y_i) \\ &= \sum_{i=1}^{n} k_i v_i(Y_i). \end{aligned}$$

It is clear that the risk exchange $(Y_1, \cdots, Y_n)$ is a solution to the weighting problem for the minimization of $\sum_{i=1}^n k_i \mathbb{E}[v_i(X_i)]$ and hence Pareto optimal. □

Appendix 6.B Conditional Mean Risk Sharing

Proof Recall the property of convex order from Sect. 2.3.2 that, whatever the random variable Z, $\mathbb{E}[X|Z] \preceq_{CX} X$. This implies the first result of the theorem. Then, we use the proof by contradiction to show Pareto optimality. Suppose that there exists a risk pool $g_1(S), ...g_n(S)$ such that

$$g_i(S) \preceq_{CX} \mathbb{E}[X_i|S] \quad \text{for } i = 1, ..., n, \text{ and } \sum_{i=1}^{n} g_i(S) = S$$

with at least one strict inequality. It follows from Theorem 2.4 that for some $k \in 1, ..., n$ there exists a probability level p such that

$$\text{TVaR}_p[g_k(S)] < \text{TVaR}_p[\mathbb{E}[X_k|S]].$$

Since TVaR is a subadditive risk measure, we have

$$\begin{aligned} \text{TVaR}_p[S] &\leq \sum_{i=1}^{n} \text{TVaR}_p[g_i(S)] \\ &< \sum_{i=1}^{n} \text{TVaR}_p[\mathbb{E}[X_i|S]] \\ &= \text{TVaR}_p[S]. \end{aligned}$$

The last equality follows from Theorem 2.5 using the assumption that the conditional means are comonotonic. Note that the above result clearly causes a contradiction. Thus, the conditional mean risk sharing is optimal when it results in comonotonic contributions to the pool. □

Appendix 6.C Proof of Theorem 6.3

Proof Let us first investigate the quantiles of all exchanged risks, i.e. $\text{VaR}_{\alpha_i}[Y_i]$ for $i = 1, \cdots, n$ (Fig. 6.7).

Quantile of Y_n:

It follows from (6.8) that Y_n is the mixed random variable with a probability mass $\mathbb{P}(S > \gamma_{n-1}) = \sum_{i=1}^{n-1} \hat{\alpha}_i$ at the point m and a continuous component. Its distribution function is given by

$$F_{Y_n}(y) = \begin{cases} F_S(y), & y < m; \\ F_S(y) + \sum_{i=1}^{n-1} \hat{\alpha}_i, & m \le y < \gamma_{n-1}; \\ 1, & \gamma_{n-1} \le y. \end{cases}$$

See Fig. 6.7(a) for the distribution of S. To obtain the α_n-quantile of Y_n, we set y such that $F_{Y_n}(y) = \alpha_n$. Because $m < \gamma_n$, this is equivalent to

$$F_S(y) + \sum_{i=1}^{n-1} \hat{\alpha}_i = 1 - \hat{\alpha}_n.$$

Solving the above equation for y yields that

$$y = \mathrm{VaR}_{(1-\sum_{i=1}^{n} \hat{\alpha}_i)}(S) = \gamma_n.$$

In other words, $\mathrm{VaR}_{\alpha_n}(Y_n) = \gamma_n$.
Quantile of Y_i for $i = 1, \cdots, n-1$:

It follows from (6.7) for $i \le n-1$ that Y_i's are also mixed random variables with probability masses at 0. The distribution function of Y_i is given by

$$F_{Y_i}(y) = \begin{cases} \alpha_i & y < \gamma_i - m; \\ \alpha_i + F_S(y+m) - F_S(\gamma_i), & \gamma_i - m \le y < \gamma_{i-1} - m; \\ 1, & \gamma_{i-1} - m \le y. \end{cases}$$

See Fig. 6.7(b) for distributions of Y_i's. Because $\mathbb{P}(Y_i > 0) \le \hat{\alpha}_i$, we must have $\mathrm{VaR}_{\alpha_i}(Y_i) \le 0$.

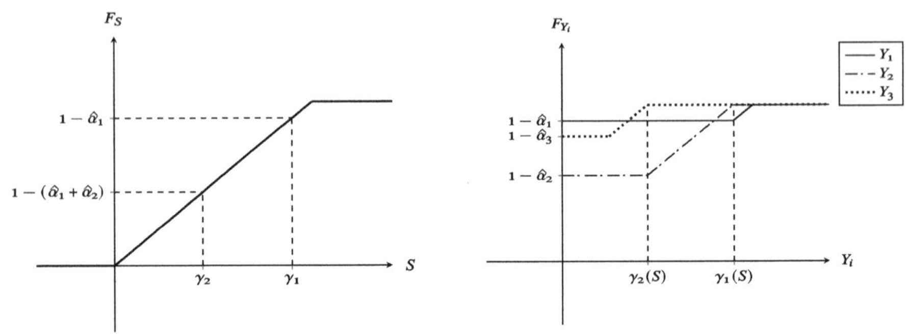

(a) Distribution of S (b) Distribution of Y_i

Fig. 6.7 Distributions in the case of $m = 0$

Therefore, it is clear that

$$\sum_{i=1}^{n} \mathrm{VaR}_{\alpha_i}(Y_i) \leq \gamma_n. \tag{6.9}$$

We can also show that, for any arbitrary risk pool $(Z_1, \cdots, Z_n)$, γ_n is in fact a lower bound for the sum of their VaRs, i.e.

$$\gamma_n \leq \sum_{i=1}^{n} \mathrm{VaR}_{\alpha_i}(Z_i). \tag{6.10}$$

While we only show the case for $n = 2$, an induction argument can be used to prove the general case for $n > 2$. The inequality $\gamma_2 \leq \mathrm{VaR}_{\alpha_1}(Y_1) + \mathrm{VaR}_{\alpha_2}(Y_2)$ where $\gamma_2 = \mathrm{VaR}_{1-\hat{\alpha}_1-\hat{\alpha}_2}[S]$ is equivalent to $F_S(s) \geq 1 - (\hat{\alpha}_1 + \hat{\alpha}_2)$, where $s = \mathrm{VaR}_{\alpha_1}[Y_1] + \mathrm{VaR}_{\alpha_2}[Y_2]$.

It is known from De Morgan's Laws that for arbitrary sets A_1 and A_2 that

$$\mathbb{P}\left(A_1 \bigcap A_2\right) \geq \mathbb{P}(A_1) + \mathbb{P}(A_2) - 1.$$

Set $A_i = \{Y_i \leq \mathrm{VaR}_{\alpha_i}[Y_i]\}$ for $i = 1, 2$. Then $\mathbb{P}(A_i) = \alpha_i$. It follows that

$$F_S(s) \geq \mathbb{P}(Y_1 \leq y_1, Y_2 \leq y_2) = \mathbb{P}\left(A_1 \bigcap A_2\right) \geq \mathbb{P}(A_1) + \mathbb{P}(A_2) - 1.$$

Thus, it is clear that $F_S(s) \geq 1 - (\hat{\alpha}_1 + \hat{\alpha}_2)$.

Combining the two inequalities (6.9) and (6.10), we conclude that the risk pool as defined in (6.7) and (6.8) is the minimizer for the objective function (6.4). □

Problems

1. Consider a three-agent Pareto risk sharing arrangement based on their respective expected utilities. Suppose that $k_i v_i' s$ are known to be

$$k_1 v_1'(y) = e^{y};$$
$$k_2 v_2'(y) = e^{y/3};$$
$$k_3 v_3'(y) = e^{y/5}.$$

Let S denote the total loss from all three agents. Show that the Pareto optimal risk exchange for the three agents is given by $y_1(S) = (1/9)S$, $y_2(S) = (1/3)S$, $y_3(S) = (5/9)S$ for $S > 0$.

2. Pareto optimal risk exchange is linear in each layer when the utility functions are exponential, as shown in Example 6.4. Similarly, Pareto optimal risk exchanges are also linear in the total loss when utility functions are power functions or log functions.

   a. Suppose that two agents have utility functions $u_i(x) = -(a_i - x)^b$ for $i = 1, 2$. The parameters are given that $a_1 = 10, a_2 = 20$, and $b = 2$. The first agent can only take limited liability up to 15, i.e. $Y_1 < 15$. Determine the fair Pareto optimal risk exchange for the two agents.
   b. Suppose that two agents have utility functions $u_i(x) = \ln(x + c_i)$ for $i = 1, 2$ where $c_1 = 10$ and $c_2 = 20$. The first agent can only take limited liability up to 15. Determine the fair Pareto optimal risk exchange.

3. Consider continuous random variables $X_1, ..., X_n$ with joint distribution F_X (probability density function f_X) and expected values $0 < \mathbb{E}[X_i] < \infty$. Define a random vector $Y^i = (Y_1^i, ..., Y_n^i)$ with a joint distribution function F_{Y^i} such that

$$F_{Y^i}(y_1, .., y_n) = \int_0^{y_1} \cdots \int_0^{y_n} \frac{x_i \, \mathrm{d}F_X(x_1, ..., x_n)}{\mathbb{E}[X_i]}$$
$$= \frac{\mathbb{E}[X_i | X_1 \leq y_1, ..., X_n \leq y_n]}{\mathbb{E}[X_i]} F_X(y_1, ..., y_n).$$

   a. Prove that, for any $s \geq 0$,

$$\mathbb{E}[X_i | S = s] = \mathbb{E}[X_i] \frac{f_{Y_1^i + ... + Y_n^i}(s)}{f_{X_1 + ... + X_n}(s)}.$$

   b. If $X_1, ..., X_n$ takes value in $\{0, 1, 2, ...\}$, then for any $s \in \{0, 1, 2, ...\}$,

$$\mathbb{E}[X_i | S = s] = \mathbb{E}[X_i] \frac{\mathbb{P}[Y_1^i + ... + Y_n^i = s]}{\mathbb{P}[X_1 + ... + X_n = s]}.$$

Chapter 7
P2P Risk Exchange

In contrast to aggregate risk pools, peer-to-peer risk exchanges are proposed as decentralized alternatives. This type of risk exchange enables a group of participants to trade risks pairwise with each other rather than relying on a facilitator, such as an insurer, takaful operator, or mutual aid platform, to handle the allocation of risks. While this topic is an area of early research, we show a few relatively simple peer-to-peer risk sharing rules, including least squares risk exchange, minimum variance risk exchange, and altruistic risk exchanges.

7.1 P2P Risk Exchange

While we have presented a variety of decentralized schemes from practice to theory in the previous sections, they are all variations of a centralized structure. Under such a structure, a facilitator is required to collect claims and allocate the losses among participants.

Such a structure is with analogy to the client-server model in the architecture of computing facilities (see Fig. 7.1). Requests from all clients for content or services are sent to a central authority as a service provider. The central node decides on how services are provided to each of its clients. Such models are used in many service industries including finance and insurance. The advantage of client-server models is that the server can achieve an economy of scale by amassing a large volume of clients and gaining expertise and efficiency in standardized services. However, it comes with a disadvantage that the functioning of the business model depends entirely on the central service provider. If the central node shuts down, the entire system collapses. The centralization of service also gives the central authority the power to deny services to clients and set terms and conditions of services in its own favor.

R. Feng, *Decentralized Insurance*, Springer Actuarial,
https://doi.org/10.1007/978-3-031-29559-1_7

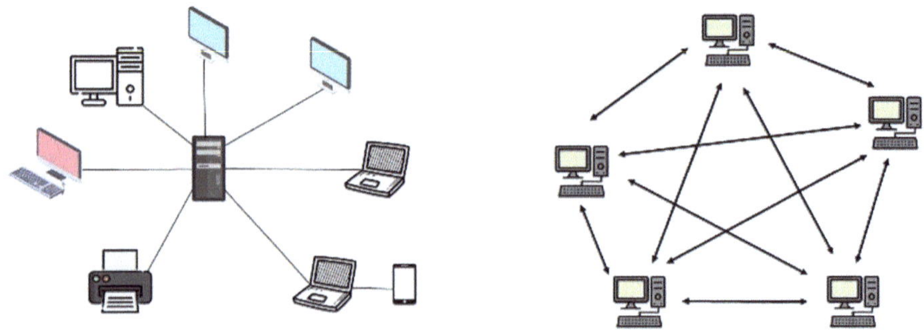

Fig. 7.1 Client-server versus Peer-to-Peer

As alternatives to decentralized models, peer-to-peer (P2P) models were developed to allow exchange services among peers. All peers are considered equipotent participants in a network. Unlike the one-to-many relationship between a server and clients, the P2P structure allows each peer to be both a server and a client. There has recently been a rise in P2P models in many financial and insurance products. For example, P2P lending platforms allow prospective borrowers to request funding for a loan from a group of creditors. The creditors search through a collection of borrowers and partially fund loans based upon their rate of return and credit worthiness among other factors. After a loan is funded, the borrower makes monthly payments into a pool that is distributed to creditors in proportion to the amount which they funded the loan. Under such models, counterparties transact directly to deliver goods and services and exchange payments without a traditional intermediary third party such as a bank. There are also recent developments in the decentralized finance space for P2P insurance models such as Teambrella and Ensuro. Investors can provide liquidity to capital pools where insurance coverage can be provided to policyholders. The basic principles have been summarized and extended to a network structure in recent works including (Abdikerimova & Feng, 2022; Charpentier et al., 2021).

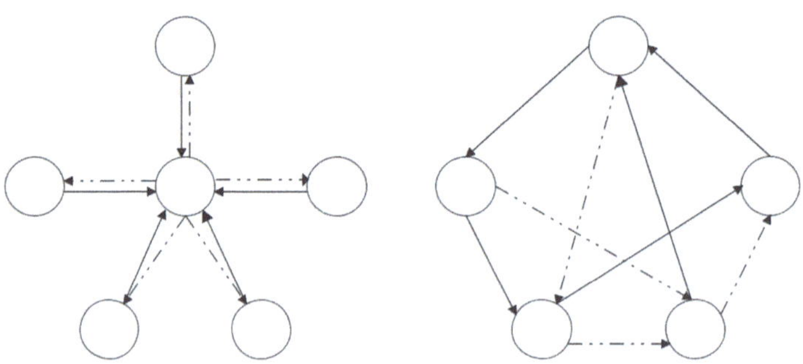

Fig. 7.2 Aggregate risk pool versus P2P risk exchange

The architecture of P2P network frameworks differs from that of aggregate risk pools. We graphically summarize such differences in Fig. 7.2. The advantage of a P2P model over the centralized model is the removal of a centralized clearing mechanism. In classical risk sharing, claims are pooled through a central entity who in turn disperses the aggregated claim among all participants. In contrast, the P2P risk sharing framework enables each claim to be paid directly from one participant to another without the process of aggregation. In other words,

$$\sum_{i=1}^{n} X_i = \sum_{i=1}^{n} Y_i, \qquad Y_i = h_i(X_1, \cdots, X_n) \text{ for } i = 1, \cdots, n, \tag{7.1}$$

for some deterministic functions h_i, $i = 1, \cdots, n$. Note that the aggregate risk sharing can be viewed as a special case where $h_i(X_1, \cdots, X_n) = h_i(S)$. In particular, we shall be investigating quota-share rules, such as $h_i(X_1, \cdots, X_n) = \sum_{j=1}^{n} \alpha_{ij} X_j$, where α_{ij} represents a proportion of peer j's risk to be carried by peer i. The main reason for such a structure is to allow transactions to take place from peer to peer without a central authority, using for example blockchain technology. As shown in Fig. 7.2, there could be cases where some nodes are disconnected, which impose additional constraints on the allocation function $\boldsymbol{h}$.

7.2 Least Squares P2P Risk Exchange

A natural question to ask is how to turn a risk pooling scheme into a risk exchange scheme. In other words, we want to avoid the process of aggregating all losses before distributing them back to individual entities.

Suppose that there is a well-defined risk pool $\boldsymbol{Y} = (Y_1, \cdots, Y_n)$ where $Y_i = h_i(S)$. The goal is to devise a peer-to-peer risk exchange $\{Y_{ij}, i, j = 1, \cdots, n\}$ where Y_{ij} represents the risk transfer from agent i to agent j. It is desirable that the risk exchange is consistent with the risk pool, i.e. $\sum_{j=1}^{n} Y_{ji} = Y_i$ for all $i = 1, \cdots, n$. While there are infinitely many ways to come up with a risk exchange scheme, we want to minimize redundancy in cash flows.

To best illustrate the algorithm, we can start with a two-agent case and then extend it to the multi-agent case. Because all losses are represented in monetary amount in practice, we shall avoid unnecessary cash flows. For example, if agent 1 pays \$1 to agent 2 and receives \$2 from agent \$2, it requires less cash transfer for agent 2 only to pay \$1 to agent 1. Therefore, to avoid redundancy in cash flows, it suffices to document either Y_{12} or Y_{21}. For the ease of presentation, we shall keep both of them but impose the symmetric payment condition that $Y_{12} = -Y_{21}$, which achieves the same purpose. In other words, the amount one agent receives from the other is the same as the amount the other agent owes to the agent. To lower the transaction cost, an agent should not transfer cash to itself, i.e. $Y_{22} = Y_{22} = 0$. Therefore, there is only one variable to be determined, namely Y_{12}. To make the risk exchange consistent with the risk pool (Y_1, Y_2), we require that the net cost for each agent to be

$$Y_1 = X_1 + Y_{21};$$
$$Y_2 = X_2 + Y_{12}.$$

It is clear that the symmetric payment $Y_{12} = -Y_{21}$ is equivalent to the market clearing condition $X_1 + X_2 = Y_1 + Y_2$. It is easy to show that the solution is given by

$$Y_{21} = Y_1 - X_1.$$

While there is a one-to-one correspondence between the risk pool and the risk exchange in the case of two agents, it is not true in general when there are more than two agents. Nonetheless, we can minimize the overall cash flows in the network by minimizing the least squares

$$\sum_{i=1}^{n} \sum_{j>i}^{n} Y_{ij}^2,$$

subject to the consistency condition with the risk pool, for all $i = 1, \cdots .n$,

$$Y_i - X_i = \sum_{j=1, j \neq i}^{n} Y_{ji}.$$

The least squares objective allows us to minimize the transaction costs of cash flows in a P2P payment system. We assume that all nodes in the P2P network are connected. In other words, there could be an exchange of risks between any pair of nodes. Represented in a matrix form, the least squares risk exchange is given by

$$\hat{\boldsymbol{Z}} = \arg\min \operatorname{tr}(\boldsymbol{Z}\boldsymbol{Z}^\top)$$

where $\boldsymbol{Z}$ is an upper triangle matrix given by

$$\boldsymbol{Z} = \begin{pmatrix} 0 & Y_{21} & \cdots & Y_{n1} \\ 0 & 0 & \ddots & \vdots \\ \vdots & \ddots & \ddots & Y_{n,n-1} \\ 0 & \cdots & 0 & 0 \end{pmatrix},$$

subject to the consistency condition

$$(\boldsymbol{Z} - \boldsymbol{Z}^\top)\boldsymbol{e} = \boldsymbol{Y} - \boldsymbol{X}, \tag{7.2}$$

where $\boldsymbol{e}$ is a column vector of one's of appropriate dimension.

Theorem 7.1 *Given a risk pool $\boldsymbol{Y}$ from the original risk vector $\boldsymbol{X}$, the corresponding least squares risk exchange is given for $i, j = 1, \cdots, n, i < j$ by*

$$Y_{ij} = \frac{1}{n}\Big[Y_j - X_j - (Y_i - X_i)\Big]. \tag{7.3}$$

The least squares risk exchange can be written in matrix form as $(1/n)[(\boldsymbol{X} - \boldsymbol{Y}) \otimes \boldsymbol{e}^\top - (\boldsymbol{X} - \boldsymbol{Y})^\top \otimes \boldsymbol{e}]$ where $\otimes$ is the Kronecker product.

If $\boldsymbol{A} = (a_{ij})$ is $m \times n$ matrix and $\boldsymbol{B}$ is $p \times q$ matrix, then the Kronecker product is a $mp \times nq$ matrix with blocks $\boldsymbol{A} \otimes \boldsymbol{B} = (a_{ij}\boldsymbol{B})$.

Note that such a risk exchange may not be linear in general as it depends on the corresponding risk pool. The least squares risk exchange suggests that each agent is to split its net flow of risk from the exchange $Y_i - X_i$ equally into n parts and attribute one part to each of the other participants. Therefore, the transfer of risk from agent i to agent j is the net of the cost carried over from agent j, $(Y_j - X_j)/n$, and the cost to pass on to agent j, $(Y_i - X_i)/n$. The proof is relegated to Appendix 7.A.

Example 7.1 Consider the unconstrained Pareto optimal risk sharing with exponential utility in Example 6.1. It is known that, given α_i, μ_i, and $\alpha = 1/\sum_i(1/\alpha_i)$, the Pareto optimal risk sharing rule is given by an aggregate risk sharing rule,

$$Y_i = \mu_i + \frac{\alpha}{\alpha_i}(S - \mu_S)$$

where $S = \sum_i X_i = \sum_i Y_i$ and $\mu_S = \sum_i \mu_i$. We can in fact turn this aggregate risk sharing into a P2P risk sharing rule by Theorem 7.1. Thus, the risk exchange between peers i and j in a complete network can be given by

$$\begin{aligned} Y_{ij} &= \frac{1}{n}\Big[\mu_j + \frac{\alpha}{\alpha_j}(S - \mu_S) - X_j - (\mu_i + \frac{\alpha}{\alpha_i}(S - \mu_S) - X_i)\Big] \\ &= \frac{1}{n}\Big[\mu_j - \mu_i + \alpha\left(\frac{1}{\alpha_j} - \frac{1}{\alpha_i}\right)(S - \mu_S) - (X_j - X_i)\Big]. \end{aligned}$$

7.3 Minimum Variance Risk Exchange

Quota-share is one of the most commonly used risk sharing rules in practice. Quota-share risk exchange already exists in today's market practice. For example, a primary insurer acquires risks from a large client base and the insurer chooses to pass on a portion of its risks to a reinsurer or a group of reinsurers through quota-share treatises.

However, most quota-share risk sharing treatises in practice are one way, in the sense that one party takes on some risk from another party. In the following, we shall focus on risk exchanges where any pair of parties trade risks in both directions.

7.3.1 Pareto Optimal Quota-Share

Let us first consider a risk exchange between two economic agents and then extend it to an arbitrary number of agents later. Assume that the potential losses of two agents are given by X_1 and X_2, respectively, with finite mean and variance. The two agents seek to engage in a risk exchange under which each agent takes on a portion of risk from the other. It is shown in Sect. 3.3.4 that, if the party ceding some risk wishes to retain a variance of fixed size and the party accepting the risk intends to minimize its variance, then a quota sharing treaty offers the most desirable result. We shall construct quota-share risk exchange.

Denote by α_{ij} the proportion of agent j's loss to be paid by agent i. From the viewpoint of agent j, her loss has to be absorbed by the group and the matter of concern is the allocation of loss to group members. Hence, we shall also refer to these proportions as allocation ratios. Then all the proportions can be represented in an allocation ratio matrix denoted by

$$\boldsymbol{A} = \begin{pmatrix} \alpha_{11} & \alpha_{12} \\ \alpha_{21} & \alpha_{22} \end{pmatrix}.$$

Denote by Y_1 and Y_2 the post-exchange risks for the two agents. According to the quota-share rule, each member carries a portion of risk from the other party. In other words, $Y_1 = \alpha_{11}X_1 + \alpha_{12}X_2$ and $Y_2 = \alpha_{21}X_1 + \alpha_{22}X_2$. In a matrix form, the post-exchange risks are a linear transformation of the pre-exchange risks, i.e. $\boldsymbol{Y} = \boldsymbol{A}\boldsymbol{X}$, where $\boldsymbol{Y} = (Y_1, Y_2)^\top$, $\boldsymbol{X} = (X_1, X_2)^\top$. In the case of P2P risk exchange, the loss conservation condition (market clearing condition)

$$\sum_i X_i = \sum_i Y_i$$

implies the conditions on the allocation coefficients

$$\alpha_{11} + \alpha_{21} = 1, \qquad \alpha_{12} + \alpha_{22} = 1.$$

The conservation condition is to ensure that all losses are completely covered by the exchange. In this section, we shall use the variance to establish a preference order. The post-exchange risk Y is preferred over the pre-exchange risk X, $Y \succ X$ if the variance of Y is smaller than that of X, i.e. $\mathbb{V}[Y] < \mathbb{V}[X]$. We assume that both agents aim to reduce as much as possible their variances by participating in a risk exchange.

Observe that if (X_1, X_2) is exchangeable (both have the same marginal distribution), then

$$\begin{aligned}\mathbb{V}[Y_1] &= \alpha_{11}^2\mathbb{V}[X_1] + \alpha_{12}^2\mathbb{V}[X_2] + 2\alpha_{11}\alpha_{12}\rho[X_1, X_2]\sqrt{\mathbb{V}[X_1]}\sqrt{\mathbb{V}[X_2]} \\ &\leq (\alpha_{11} + \alpha_{12})^2\mathbb{V}[X_1] = \mathbb{V}[X_1].\end{aligned}$$

Similarly, we can also show that $\mathbb{V}[Y_2] \leq \mathbb{V}[X_2]$. Therefore, the two agents are always better off by engaging in a risk exchange. Note that the P2P risk exchange network does include the non-exchange situation as a special case where $\alpha_{ij} = 0$ for all $i \neq j$, which says none of the agents has any cash transfer with anybody else. Therefore, even in general cases, the individual rationality $\mathbb{V}[Y_i] \leq \mathbb{V}[X_i]$ is automatically included in the optimization problem.

7.3.2 *Fair Pareto Risk Exchange*

As the agents have competing interests to reduce their respective variances, we aim for a Pareto optimality for both agents.

As discussed in Sect. 3.4, such a Pareto optimal solution is equivalent to solving the following optimization problem:

$$\min_{A} \{w_1\mathbb{V}[Y_1] + w_2\mathbb{V}[Y_2]\}, \tag{7.4}$$

for arbitrary weights $w_1, w_2 \geq 0$, subject to the conservation constraints

$$\alpha_{11} + \alpha_{21} = 1, \qquad \alpha_{12} + \alpha_{22} = 1.$$

Note that the fairness condition is not yet included as a constraint for this optimization problem at this point.

We represent post-exchange variances graphically in Fig. 7.3 to illustrate how Pareto solutions are found. Note that post-exchange variances are given by

$$\begin{aligned}\mathbb{V}(Y_1) &= \alpha_{11}^2\sigma_1^2 + (1-\alpha_{22})^2\sigma_2^2 + 2\alpha_{11}(1-\alpha_{22})\rho\sigma_1\sigma_2, \\ \mathbb{V}(Y_2) &= (1-\alpha_{11})^2\sigma_1^2 + \alpha_{22}^2\sigma_2^2 + 2\alpha_{22}(1-\alpha_{11})\rho\sigma_1\sigma_2,\end{aligned}$$

where σ_1, σ_2 are standard deviations of X_1 and X_2, respectively, and ρ is the correlation coefficient between X_1 and X_2. Note that X_1 and X_2 can have different distributions, not necessarily normal. Then, the contours of constant variance for the two agents are represented by ellipses in the $(\alpha_{11}, \alpha_{22})$-plane.

Any given point in the $(\alpha_{11}, \alpha_{22})$-plane passes through one and only one ellipse of the family E_1 and another ellipse of the family E_2. Consider a point that moves along the ellipse E_1^* with constant variance $\mathbb{V}(X_1)$. As this point changes position,

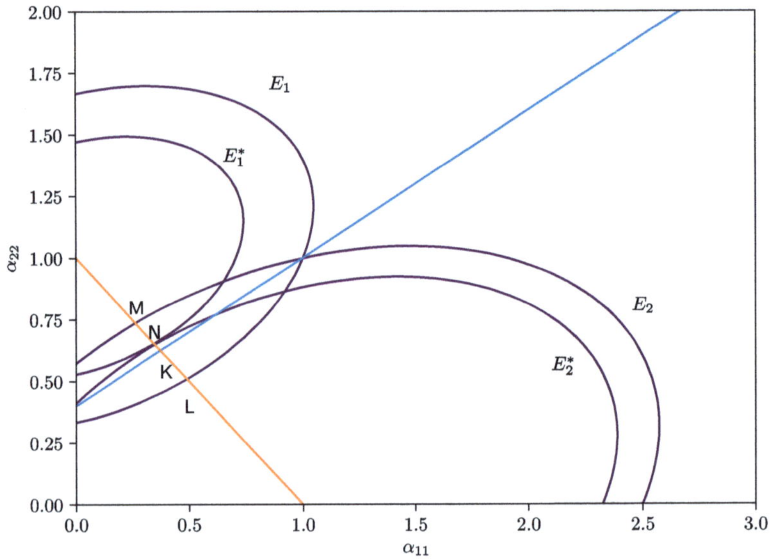

Fig. 7.3 Illustration of fair Pareto P2P risk exchange. $\mu_1 = 3, \mu_2 = 5, \sigma_1 = 2, \sigma_2 = 3$, and $\rho = 0.3$

the corresponding $\mathbb{V}(X_2)$ varies and obtains the minimum value when the location reaches the point N at which the ellipse E_1^* is tangential to another ellipse E_2^* of the E_2 family. One can show that the set of all Pareto optimal points lie within the line segment ML represented by

$$\text{(Pareto optimality)} \qquad \alpha_{11} + \alpha_{22} = 1.$$

In order to determine a unique Pareto optimal solution, it is further stipulated that $\mathbb{E}(X_i) = \mathbb{E}(Y_i)$. This corresponds to the actuarial fairness condition in the statement of this problem, which means that each member's expected income matches the expected outgo, i.e.

$$\text{(actuarial fairness)} \qquad \alpha_{11}\mu_1 + (1 - \alpha_{22})\mu_2 = \mu_1.$$

The set of all allocation ratios that meet the fairness condition is represented by the blue line. Note that the two straight lines intersect with each other at the point K, which represents the fair Pareto optimal solution. Combining the optimality and the actuarial fairness conditions yields the solution

$$\alpha_{11} = \frac{\mu_1}{\mu_1 + \mu_2}, \quad \alpha_{22} = \frac{\mu_2}{\mu_1 + \mu_2}.$$

Note that the fairness condition can be applied with premium principles other than the expected value principle.

Bear in mind that the fairness condition is not included in the optimization problem itself. There is no guarantee that the solution lies in the interior of the intersection of the two ellipses E_1 and E_2. One can show that a fair Pareto optimal risk sharing exists if and only if

$$\frac{\mu_1^2}{(\mu_1+\mu_2)^2} \le \frac{\sigma_1^2}{\sigma_1^2+\sigma_2^2+2\rho\sigma_1\sigma_2}$$
$$\frac{\mu_2^2}{(\mu_1+\mu_2)^2} \le \frac{\sigma_2^2}{\sigma_1^2+\sigma_2^2+2\rho\sigma_1\sigma_2}.$$

When such conditions fail, this problem does not deliver any solution that satisfies both the fairness and Pareto optimality.

Observe that the fair Pareto optimality for P2P risk exchange could be extended to a general case of arbitrary dimension. There are n economic agents who want to reduce their respective variances by engaging in a P2P risk exchange. Denote the pre-exchange risk vector of the n agents by $\boldsymbol{X} = (X_1, \cdots, X_n)$, and the post-exchange risk vector by $\boldsymbol{Y} = (Y_1, \cdots, Y_n)$. Suppose that $\boldsymbol{X}$ has finite mean $\mathbb{E}(\boldsymbol{X}) = \boldsymbol{\mu}$ and positive definite covariance matrix $\mathbb{C}(\boldsymbol{X}) = \boldsymbol{\Sigma}$. Since we are interested in quota-share P2P risk exchanges, it is assumed that the post-exchange risk vector is a linear transformation of the pre-exchange risk vector, i.e. $\boldsymbol{Y} = \boldsymbol{A}\boldsymbol{X}$, where $\boldsymbol{A} = (\alpha_{ij})_{i,j=1,\cdots,n}$ is the allocation coefficient matrix for the n-dimensional risk exchange. The search for a Pareto optimal P2P risk exchange boils down to the optimization problem with respect to an allocation coefficient matrix,

$$\hat{\boldsymbol{A}} = \arg\min_{\boldsymbol{A}} \sum_{i=1}^{n} \omega_i \mathbb{V}(Y_i) = \arg\min_{\boldsymbol{A}} \operatorname{tr}(\boldsymbol{W}\boldsymbol{A}\boldsymbol{\Sigma}\boldsymbol{A}^\top), \tag{7.5}$$

where $\boldsymbol{W}$ is a diagonal matrix with arbitrary weight $\omega_i > 0$ in the i-th place on its main diagonal. The operation arg min refers to the process of finding an argument of minima, or an argument that minimizes the objective function. Recall the loss conservation constraints

$$\boldsymbol{e}^\top \boldsymbol{A} = \boldsymbol{e}^\top$$

where the column vector $\boldsymbol{e}$ consists of all ones. Observe that (7.5) is an n-dimensional extension of the two-dimensional problem (7.4).

Theorem 7.2 *Pareto optimal quota-share risk exchanges are given by*

$$\hat{\boldsymbol{A}} = \frac{1}{\boldsymbol{e}^\top \boldsymbol{W}^{-1}\boldsymbol{e}} \boldsymbol{W}^{-1}\boldsymbol{e}\boldsymbol{e}^\top,$$

where $\boldsymbol{W}$ is a diagonal matrix with arbitrary positive constants on its main diagonal.

Note that this optimal risk exchange is *non-olet*. The allocation coefficients differ by who pays for the loss. In other words, the origin of the loss does not matter, i.e. $\hat{\boldsymbol{A}} = (\alpha_{i,j}), \alpha_{i,j} = \alpha_i$ for all $j = 1, \ldots, n$. Another important fact about this risk exchange is that the allocation does not depend on the probability distribution of underlying risks. As we have seen in Chap. 6, this is a common feature of Pareto optimal aggregate (non-olet) risk sharing. The optimal quota-share allocation coefficients are in proportion to the agents' harmonic weights representing their "significance" in the decision-making,

$$(\alpha_1, \ldots, \alpha_i, \ldots \alpha_n) = \left(\frac{1/\omega_1}{\sum_{j=1}^n 1/\omega_j}, \ldots, \frac{1/\omega_i}{\sum_{j=1}^n 1/\omega_j}, \ldots, \frac{1/\omega_n}{\sum_{j=1}^n 1/\omega_j} \right).$$

If the weight ω_i is large in the optimization objective $\sum_i \omega_i \mathbb{V}[Y_i]$, the variance of agent i's cost should be kept small. This is achieved by giving agent i a small portion of everyone else's loss. This explains why the allocation coefficients are in proportion to *harmonic weights* $(1/\omega_i)/(\sum_{j=1}^n 1/\omega_j)$. If all participants are given equal weight, i.e. $\omega_1 = \cdots = \omega_n$, then the risk sharing allocation becomes uniform and all participants carry an equal amount of the losses of each other, i.e. $\alpha_1 = \cdots = \alpha_n = 1/n$.

Another approach to identify a unique risk exchange is to consider the actuarial fairness condition, $\mathbb{E}[\boldsymbol{X}] = \mathbb{E}[\boldsymbol{Y}]$, which implies that

$$\boldsymbol{A\mu} = \boldsymbol{\mu}.$$

Theorem 7.3 *The fair Pareto optimal quota-share risk exchange is determined by*

$$\frac{1}{\boldsymbol{e}^\top \boldsymbol{M e}} \boldsymbol{M e e}^\top,$$

where $\boldsymbol{M}$ *is a diagonal matrix with elements of* $\boldsymbol{\mu}$ *on its main diagonal.*

Under the actuarial fairness condition, the risk exchange is again non-olet. Observe that $\boldsymbol{A} = (\alpha_{i,j}), \alpha_{i,j} = \alpha_i$ for all $i, j = 1, \ldots, n$ and

$$(\alpha_1, \ldots, \alpha_n) = \left(\frac{\mu_1}{\sum_{j=1}^n \mu_j}, \ldots, \frac{\mu_n}{\sum_{j=1}^n \mu_j} \right).$$

Regardless of the source of the loss, each participant carries a portion of the loss in proportion to her expected loss. Here we can think of an individual with a high expected loss as a high-risk individual. This result clearly agrees with the expectation that a high-risk individual should be asked to pay more than a low-risk peer.

The proofs of both theorems in this section can be found in Appendix 7.B.

7.3.3 *Pareto in the Class of Fair Risk Exchanges*

A fair Pareto optimal P2P risk exchange is first of all a Pareto optimal solution for all participating agents that is also actuarially fair. As alluded to earlier, such a solution may not always exist. If the actuarial fairness is critical, then we present an alternative approach to P2P risk exchange by which only fair risk exchanges are considered. In other words, we search through the narrower class of fair risk exchanges to obtain Pareto optimality.

Take the two-agent case as an example again. As before, the Pareto optimality can be found through the minimization of a single objective function that is a weighted average of both parties' objectives. We consider the following optimization problem:

$$\min_{A} \{w_1 \mathbb{V}(Y_1) + w_2 \mathbb{V}(Y_2)\},$$

for arbitrary weights $w_1, w_2 \geq 0$, subject to the constraints

$$\text{(zero-balance conservation)} \qquad \alpha_{11} + \alpha_{21} = 1, \qquad \alpha_{12} + \alpha_{22} = 1.$$

$$\text{(actuarial fairness)} \qquad \alpha_{11}\mu_1 + (1 - \alpha_{22})\mu_2 = \mu_1.$$

In contrast to the previous section, the actuarial fairness is now a constraint to the optimization problem, rather than an extra condition to pin down a unique solution.

Let us visualize the process of finding a Pareto optimal solution in Fig. 7.4. In the top panel, the contours of constant variances for the two agents are shown in the $(\alpha_{11}, \alpha_{22})$-plane. The straight line represents all combinations of allocation for fair risk exchanges. Because we are only interested in fair risk changes, it is clear that Pareto optimal solutions must be points on the straight line in the interior of the intersection of the two ellipses, say the line segment AE. In the bottom panel, we draw the corresponding points on the $(\mathbb{V}[Y_1], \mathbb{V}[Y_2])$-plane. It is clear from this graph that neither of points A, B, E is Pareto optimal. Both agents can achieve lower variances by moving to the left and below points A, B, E along the fairness line. The points of interest are in the segment CD. If the combination of allocation coefficients moves along the fairness line in any direction, at least one of the two agents would be worse off with a higher variance. Therefore, all points in the segment CD are Pareto optimal.

We can identify a desirable P2P risk sharing rule by specifying the appropriate weights w_1 and w_2. For example, without loss of generality, we restrict the attention to the case where the variance of each agent in the pool is treated equally, namely $w_1 = 1$, $w_2 = 1$. For simplicity, we further assume that X_1 and X_2 are independent of one other. Then, the optimal solution can be solved by a quadratic equation and hence is given by

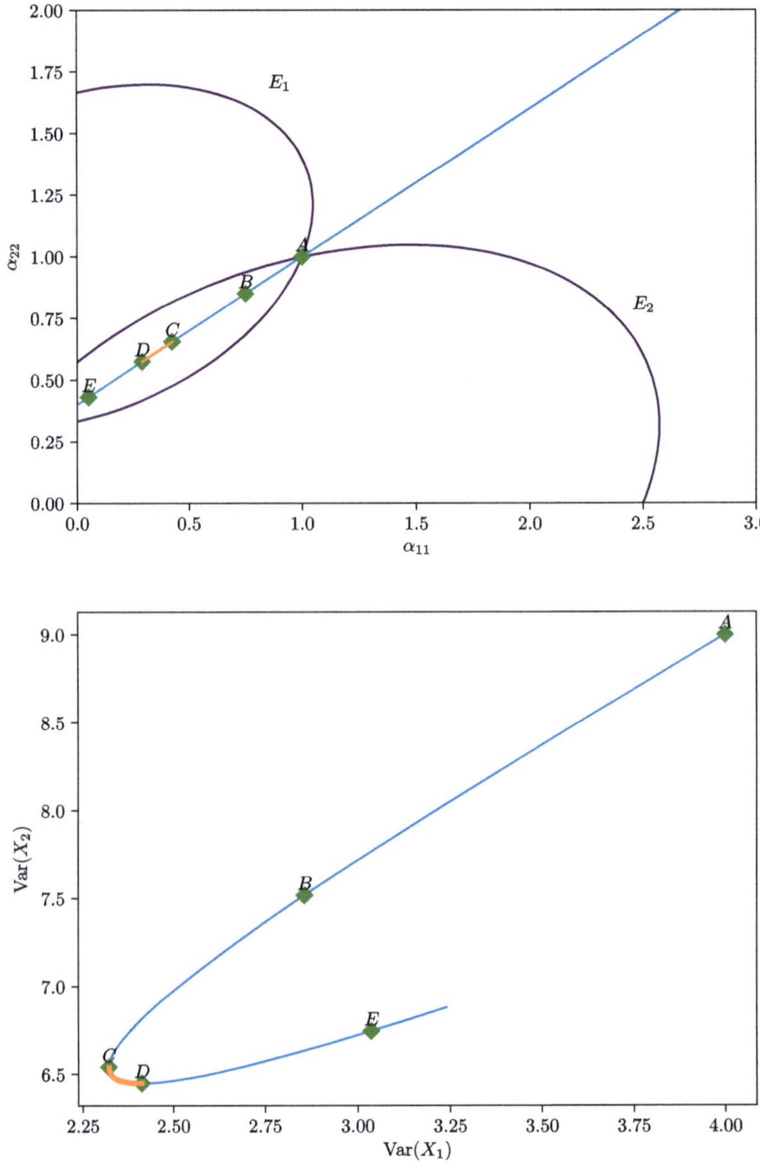

Fig. 7.4 Pareto in the class of fair risk exchanges

$$\alpha_{11} = \frac{\mu_2^2\sigma_1^2 + 2\mu_1^2\sigma_2^2 - \sigma_2^2\mu_1\mu_2}{2(\mu_2^2\sigma_1^2 + \mu_1^2\sigma_2^2)}, \quad \alpha_{22} = \frac{\mu_1^2\sigma_2^2 + 2\mu_2^2\sigma_1^2 - \sigma_1^2\mu_1\mu_2}{2(\mu_2^2\sigma_1^2 + \mu_1^2\sigma_2^2)}. \tag{7.6}$$

The discussion can be extended further to n-dimensional risk exchanges. The risk vector $\boldsymbol{X}$ represents the pre-exchange risks of n economic agents. Assume that the vector has mean $\mathbb{E}(\boldsymbol{X}) = \boldsymbol{\mu}$ and positive definite covariance matrix $\mathbb{C}(\boldsymbol{X}) = \boldsymbol{\Sigma}$. The post-exchange risk vector $\boldsymbol{Y}$ is given by a linear transformation of the pre-exchange risk vector $\boldsymbol{X}$, i.e. $\boldsymbol{Y} = \boldsymbol{A}\boldsymbol{X}$. The main optimization problem becomes

$$\hat{\boldsymbol{A}} = \underset{\boldsymbol{A}}{\text{amin}} \sum_{i=1}^{n} \mathbb{V}(Y_i) = \underset{\boldsymbol{A}}{\text{amin}} \ \text{tr}(\boldsymbol{A}\boldsymbol{\Sigma}\boldsymbol{A}^\top),$$

where we consider the optimization subject to the actuarial fairness and the loss conservation constraints

$$\boldsymbol{A}\boldsymbol{\mu} = \boldsymbol{\mu}, \qquad \boldsymbol{e}^\top \boldsymbol{A} = \boldsymbol{e}^\top.$$

The solution to this problem is explicitly known and its proof can be found in Appendix 7.C.

Theorem 7.4 *The Pareto optimal solution in the class of fair P2P risk exchanges is given by*

$$\hat{\boldsymbol{A}} = \frac{1}{n}\boldsymbol{e}\boldsymbol{e}^\top + k\left(\boldsymbol{I} - \frac{1}{n}\boldsymbol{e}\boldsymbol{e}^\top\right)\boldsymbol{\mu}\boldsymbol{\mu}^\top\boldsymbol{\Sigma}^{-1},$$

where the constant $k^{-1} = \boldsymbol{\mu}^\top\boldsymbol{\Sigma}^{-1}\boldsymbol{\mu}$ *and* $\boldsymbol{I}$ *is an n-by-n identify matrix.*

There are a number of special cases that are worth pointing out. When the means of losses from all agents are the same, i.e. $\mu_1 = \cdots = \mu_n$, then the optimal allocation matrix reduces to $\hat{\boldsymbol{A}} = (1/n)\boldsymbol{e}\boldsymbol{e}^\top$. This result suggests that the allocation of losses should be uniform as long as the means are the same. Furthermore, if both means and variances for all agents are the same, i.e. $\mu_1 = \cdots = \mu_n$ and $\sigma_1 = \cdots = \sigma_n$, we can show that each agent can observe variance reduction,

$$\mathbb{V}(Y_i) \le \mathbb{V}(X_i).$$

Observe that

$$\mathbb{V}(Y_i) = \frac{1}{n^2}\boldsymbol{e}^\top\boldsymbol{\Sigma}\boldsymbol{e} + k\mu_i^2 - \frac{k}{n^2}\left(\sum_{i=1}^{n}\mu_i\right)^2 \ge \frac{1}{n^2}\boldsymbol{e}^\top\boldsymbol{\Sigma}\boldsymbol{e},$$

where the inequality results from the Cauchy-Schwartz inequality $\sum_i \mu_i^2 - n^{-2}$ $(\sum_i \mu_i)^2 \ge 0$. Note that when the means are identical, all agents carry an equal share of the aggregate risk and hence the post-exchange variance reaches its minimum,

$$\mathbb{V}(Y_i) = \mathbb{V}\left(\sum_{i=1}^{n} X_i/n\right) = (1/n^2)\boldsymbol{e}^\top\boldsymbol{\Sigma}\boldsymbol{e}.$$

In the event that all agents have the same mean and variance, i.e. $\mu_1 = \cdots = \mu_n$ and $\sigma_1 = \cdots = \sigma_n = \sigma$, each agent observes some variance reduction, i.e.

$$\mathbb{V}(Y_i) = \sigma^2 \frac{1}{n^2} \sum_{i=1}^{n} \sum_{j=1}^{n} \rho_{ij} \leq \sigma^2 = \mathbb{V}(X_i).$$

If all risks are perfectly positively correlated, i.e. $\rho_{ij} = 1$ for all $i, j = 1, \ldots, n$, then there is no reduction in variance. There is no risk reduction from the diversification of perfectly correlated risks in the pool. Keep in mind, however, that not all risks with the same mean and variance can be made perfectly corrected, as shown in Sect. 2.5.1.

The study of Pareto optimal quota-share risk exchanges dates back to Beard et al. (1969) and Bühlmann (1980). Early discussions focus on fair Pareto optimal risk exchanges. The extensions to P2P risk sharing are introduced in Abdikerimova and Feng (2022), Feng et al. (2023), and Charpentier et al. (2021). A multi-period version of P2P risk sharing has been studied in Abdikerimova et al. (2022).

It should also be pointed out that, in a very large risk pool, it may be costly to execute direct transactions between all agents. It is also computationally challenging to determine the allocation matrix in high-dimensional cases. Feng et al. (2023) propose the concept of hierarchical risk sharing, where agents are grouped together and transactions occur between members of a group or between the groups themselves. A concept of collaborative insurance in a similar hierarchical network structure can be found in Charpentier et al. (2021).

7.4 Altruistic Risk Exchange

While the majority of risk sharing rules studied in the academic literature are based on Pareto efficiency, there are other exceptional rules. One of such rules is the so-called altruistic risk exchange. Altruism refers to the acts out of selfless concerns for the well-being of others. The practice of altruism has long history in all civilizations. Its philosophy underlies the formation of mutual support societies that perform many of the risk management functions as we have discussed in earlier chapters. Figure 7.5 is a famous painting by a French painter Jacques-Louis David in 1780. It depicts an act of kindness in the Roman time where a women offers alms to a beggar with his husband, the soldier behind her, realizing that the beggar was in fact his former commander. It is a story about communal support where people pay back to those who helped them in the past.

The concept of altruism is embedded in the practice of risk sharing. In the context of P2P risk sharing, we formulate a risk sharing rule that reflects the principle of altruism. Assume that participants of the network are interested in maximizing their own allocations of losses from other peers in order to provide the most communal

Fig. 7.5 Altruism in Ancient Rome—"Belisarius Begging for Alms" by Jacques-Louis David. *Source* http://www.warontherocks.com/

support. As we shall show later, an interesting consequence of the altruistic risk exchange is that claimants do not have to pay for their own losses. This is tied to the common philosophy that the ones who help others shall be rewarded. Altruistic risk sharing has been widely used in practice. For example, in most online mutual aid models, a legitimate claimant receives a benefit payment, the cost of which is carried by all mutual aid members other than the claimant. Such a rule is both practically convenient and can be theoretically justified with the concept of altruism.

The problem can be formulated in the same setup as in the previous section. Suppose that there are n economic agents engaging in a risk exchange. The pre-exchange risks are denoted by $\boldsymbol{X} = (X_1, \cdots, X_n)$. We aim for a quota-share risk exchange under which peer i commits to take on a portion of peer j's potential loss. Denote by α_{ij} the proportion of peer j's loss to be paid by peer i. Therefore, the post-exchange risks denoted by $\boldsymbol{Y} = (Y_1, \cdots, Y_n)$ are generated from a linear transformation of the pre-exchange risks, i.e. $\boldsymbol{Y} = \boldsymbol{A}\boldsymbol{X}$, where $\boldsymbol{A}$ is the allocation matrix $\boldsymbol{A} = (\alpha_{ij})_{i,j=1,\cdots,n}$. As shown before, the allocation matrix is expected to meet the loss conservation and actuarial fairness conditions, i.e.

$$\boldsymbol{e}^\top \boldsymbol{A} = \boldsymbol{e}^\top, \qquad \boldsymbol{A}\boldsymbol{\mu} = \boldsymbol{\mu}.$$

A quantitative measure of the degree to which peers support each other is to observe the maximal amount of cash transfers. The higher the amount of cash flow, the more peers collectively support each other. Therefore, we seek to maximize the sum of square roots of the allocation ratios.

$$\max_{\boldsymbol{A}} \sum_{i=1}^{n} \sum_{j=1, j\neq i}^{n} \alpha_{ij}^{1/2}.$$

The choice of the power $1/2$ is used to make the objective concave and hence to produce a unique solution. One could replace the power $1/2$ with any $0 < p < 1$.

Example 7.2 (*Homogeneous Case*) When loss distributions of all peers are the same, the group is considered *homogeneous*. In this case, the mean vector $\boldsymbol{\mu} = \mu \boldsymbol{e}$ for a common mean μ. Thus, the actuarial fairness equation $\boldsymbol{A}\boldsymbol{\mu} = \boldsymbol{\mu}$ reduces to

$$\boldsymbol{A}\boldsymbol{e} = \boldsymbol{e}.$$

The implication is that the matrix $\boldsymbol{A}$ is doubly stochastic. In a homogeneous case, the risks are indistinguishable. When a peer makes a claim, the split of cost should be identical for all peers. Hence, we only need to identify two coefficients, namely the self-retained portion, α^*, and the portion carried by one another, α. Let S_j^- be the sum of claims from all peers in the group other than peer j. The financial return of peer j at the end of the exchange period is $R_j = (1 - \alpha^*)X_j - \alpha S_j^-$. The actuarial fairness condition $\mathbb{E}[R_j] = 0$ implies that $(1 - \alpha^*)\mathbb{E}[X_j] = \alpha \mathbb{E}[S_j^-]$, which is the same as the loss conservation condition $(n - 1)\alpha + \alpha^* = 1$. The purpose of the altruistic plan is to maximize the cash transfer among peers when losses occur, which can be represented by

$$\max \alpha^{1/2}$$

subject to

$$(n-1)\alpha + \alpha^* = 1, \qquad 0 \le \alpha, \alpha^* \le 1.$$

It is straightforward that the solution is given by

$$\alpha^* = 0, \qquad \alpha = \frac{1}{n-1}.$$

This plan is altruistic in nature, as a claimant is not required to pay anything for her own benefit and all the cost is to be split among others.

Example 7.3 (*Heterogeneous Case*) Consider a heterogeneous group with two distinct risk classes of n_1 and n_2 participants, respectively. Figure 7.6 offers an illustration of a risk exchange among peers in a two-risk-class group. Because each risk class is a homogeneous group on its own, we treat each peer within the same risk class in exactly the same way. Therefore, we can use the simplified notation for the allocation ratios. For $i = 1, 2$, let α_{ii}^* be the self-retained proportion of a member's own loss, α_{ii} be the proportion of the loss that a member allocates to each peer other than himself/herself in the same risk class, and α_{ij} represent the proportion of loss incurred by any peer in the risk class j to be allocated to any peer in the risk class i. In other words, α_{ii} is the **intra-class allocation ratio**, while α_{ij} is the **inter-class allocation ratio.**

Consider the position of an arbitrary peer j in the class i. Let S_i^- be the sum of claims in the class i other than peer j. Let S_l be the sum of claims in class l. It follows immediately from the actuarial fairness condition that for $i, l = 1, 2$ and

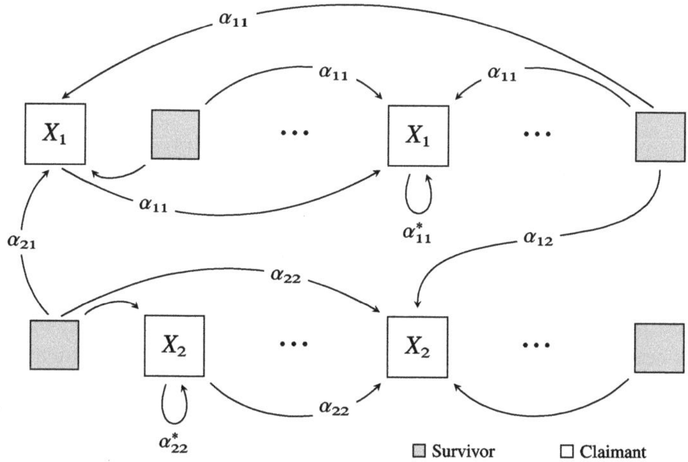

Fig. 7.6 Altruistic risk exchange between two risk classes

$i \neq l$, $\alpha_{li}\mathbb{E}[S_l] + \alpha_{ii}\mathbb{E}[S_i^-] + \alpha_{ii}^*\mathbb{E}[X_i] = \mathbb{E}[X_i]$. Let μ_l be the mean of each peer in the class l. Then the actuarial fairness condition implies that

$$\alpha_{li}n_l\mu_l + \alpha_{ii}(n_i - 1)\mu_i + \alpha_{ii}^*\mu_i = \mu_i, \quad \text{for all } i, l = 1, 2 \text{ and } i \neq l.$$

We also impose the principle of indemnity and loss conservation conditions, i.e.

$$n_l\alpha_{li} + (n_i - 1)\alpha_{ii} + \alpha_{ii}^* = 1, \quad \text{for all } i, l = 1, 2 \text{ and } i \neq l. \tag{7.7}$$

$$0 \leq \alpha_{ij}, \alpha_{ii}, \alpha_{ii}^* \leq 1.$$

We seek to maximize the sum of square roots of allocation ratios. Note that the self-retained ratios are excluded from the maximization as the consideration is selfless. Therefore,

$$\max_{A} \sum_{i=1}^{2} \sum_{l=1, l\neq i}^{2} \alpha_{li}^{1/2}.$$

The solution to the altruistic plan is given by the following:

1. If $(n_1\mu_1)/(n_2\mu_2) \leq 1$,

$$\alpha_{11} = 0, \qquad \alpha_{11}^* = 0, \qquad \alpha_{12} = \frac{\mu_1}{\mu_2}\frac{1}{n_1}, \qquad \alpha_{21} = \frac{1}{n_2},$$

$$\alpha_{22}^* = \begin{cases} 0, & n_2 > 1; \\ 1 - \frac{\mu_1}{\mu_2}, & n_2 = 1, \end{cases} \quad \alpha_{22} = \begin{cases} \frac{1-(n_1\mu_1)/(n_2\mu_2)}{(n_2-1)}, & n_2 > 1; \\ 0, & n_2 = 1. \end{cases}$$

2. If $(n_1\mu_1)/(n_2\mu_2) > 1$,

$$\alpha_{12} = \frac{1}{n_1}, \qquad \alpha_{21} = \frac{\mu_2}{\mu_1}\frac{1}{n_2}, \qquad \alpha_{22} = 0, \qquad \alpha^*_{22} = 0.$$

$$\alpha^*_{11} = \begin{cases} 0, & n_1 > 1, \\ 1 - \frac{\mu_2}{\mu_1}, & n_1 = 1; \end{cases} \quad \alpha_{11} = \begin{cases} \frac{1-(n_2\mu_2)/(n_1\mu_1)}{n_1-1}, & n_1 > 1, \\ 0, & n_1 = 1. \end{cases}$$

Observe that the two cases are symmetric with respect to risk classes 1 and 2. Take a special case as an example. When there is only one member in risk class 2, the self-retained portion of own loss is represented by α^*_{22}, and hence, the intra-class ratio α_{22} is set to 0. Assume that $n_1\mu_1/\mu_2 \leq 1$. It means that the mean loss of the single member in class 2 outweighs the combined mean loss of the whole class 1. When a loss occurs in class 1, it should be covered entirely by the member in class 2, i.e. $\alpha_{11} = \alpha^*_{11} = 0$ and $\alpha_{21} = 1$. When a loss occurs in class 2, however, it is unfair to expect members in class 1 to cover all losses from class 2 due to their disproportionate historic mean losses. To meet the fairness condition, we require the member in class 2 to self-retain a portion $\alpha^*_{22} = 1 - \mu_1/\mu_2$ of its loss. Note that $0 \leq \mu_1/\mu_2 \leq 1/n_1$ and hence $0 \leq \alpha^*_{22} \leq 1$. The rest of the loss shall be carried by class 1 where each member takes on $\alpha_{12} = (\mu_1/\mu_2)/n_1$. Note that in this case $n_1\alpha_{12} + \alpha^*_{22} = 1$.

The solution can be read further and explained why the plan is considered altruistic and fair.

- When a loss occurs, any claimant in the low-risk class receives a compensation but does not have to pay for her own loss.
- Members of a high-risk class (with a high expected loss) are required to pay a larger portion of the other class' claim than those of a low-risk class.
- When the average costs of the two risk classes are the same, i.e. $(n_1\mu_1)/(n_2\mu_2) = 1$, each risk class' total compensation is paid entirely by the other class, i.e. $\alpha_{11} = \alpha^*_{11} = \alpha_{22} = \alpha^*_{22} = 0$.
- If their average costs are unequal (for example, $(n_1\mu_1)/(n_2\mu_2) \leq 1$), then the risk class with higher cost (class 2) is required to carry some of its own cost (α^*_{22} or $\alpha_{22} > 0$). In other words, when the costs are unbalanced, the class that receives more should pay more.
- When there is only one member in a risk class, we set the intra-class allocation ratio to be zero in order to avoid the ambiguity with the self-retained ratio.

The concept of altruistic risk exchange was studied for various types of models in Abdikerimova and Feng (2022). Although they arise from different backgrounds, it is arguable that the altruistic risk sharing has the opposite effect as the least squares risk sharing introduced in Sect. 7.2 from the viewpoint of transaction cost.

Appendix 7.A Least Squares Risk Sharing

We use the Lagrange method to solve this optimization problem. Let $\boldsymbol{\lambda} = (\lambda_1, \lambda_2, \cdots, \lambda_{k-1}, 0)^\top$ be the set of Lagrange multipliers for the constraints in the consistency condition (7.2). Note that no multiplier is assigned to the last equation in (7.2). Define

$$\mathcal{L} = tr(\boldsymbol{Z}\boldsymbol{Z}^\top) + \boldsymbol{\lambda}^\top((\boldsymbol{Z} - \boldsymbol{Z}^\top)\boldsymbol{e} - (\boldsymbol{Y} - \boldsymbol{X})).$$

This gives the following first-order conditions:

$$\frac{\mathrm{d}\mathcal{L}}{\mathrm{d}Y_{ij}} = 2Y_{ij} + \lambda_j - \lambda_i = 0, \qquad \text{for } i > j, i < n;$$

$$\frac{\mathrm{d}\mathcal{L}}{\mathrm{d}Y_{nj}} = 2Y_{nj} + \lambda_j = 0, \qquad \text{for } n > j.$$

In order to simplify the notation, we introduce

$$\boldsymbol{\Lambda} = -\frac{1}{2}\begin{pmatrix} 0 & \lambda_1 - \lambda_2 & \lambda_1 - \lambda_3 & \cdots & \lambda_1 - \lambda_{k-1} & \lambda_1 \\ 0 & 0 & \lambda_2 - \lambda_3 & \cdots & \lambda_2 - \lambda_{n-1} & \lambda_2 \\ \vdots & \vdots & \vdots & \ddots & \vdots & \vdots \\ 0 & 0 & 0 & \cdots & \lambda_{n-2} - \lambda_{n-1} & \vdots \\ 0 & 0 & 0 & & 0 & \lambda_{n-1} \\ 0 & 0 & 0 & \cdots & 0 & 0 \end{pmatrix}.$$

Then we can rewrite these conditions in a matrix form.

$$\boldsymbol{Z} = \boldsymbol{\Lambda}, \qquad (\boldsymbol{Z} - \boldsymbol{Z}^\top)\boldsymbol{e} = \boldsymbol{Y} - \boldsymbol{X}. \tag{7.8}$$

It is easy to show by substitution that (7.3) solves the system of equations in (7.8).

Appendix 7.B Pareto Optimal Quota-Share Risk Exchange

Proof of Theorem 7.2
One may determine the optimal allocation using the method of Lagrange multipliers. Define the objective function $\mathcal{L}$ by

$$\mathcal{L} = \mathrm{tr}(\boldsymbol{W}\boldsymbol{A}\boldsymbol{\Sigma}\boldsymbol{A}^\top) + (\boldsymbol{e}^\top - \boldsymbol{e}^\top\boldsymbol{A})\boldsymbol{\lambda}.$$

It follows that

$$0 = \frac{\partial \mathcal{L}}{\partial \boldsymbol{A}} = \boldsymbol{W}^\top \boldsymbol{A}\boldsymbol{\Sigma}^\top + \boldsymbol{W}\boldsymbol{A}\boldsymbol{\Sigma} - \boldsymbol{e}\boldsymbol{\lambda}^\top. \tag{7.9}$$

$$0 = \frac{\partial \mathcal{L}}{\partial \boldsymbol{\lambda}} = \boldsymbol{e}^\top - \boldsymbol{e}^\top \boldsymbol{A}. \tag{7.10}$$

Note $\boldsymbol{W}$ and $\boldsymbol{\Sigma}$ are symmetric. From (7.9), $\boldsymbol{A}$ is given by

$$\hat{\mathbf{A}} = \mathbf{W}^{-1}(\mathbf{e}\boldsymbol{\lambda}^\top)(2\boldsymbol{\Sigma})^{-1}.$$

Inserting this expression into Eq. (7.10) and solving for $\boldsymbol{\lambda}$ gives

$$\mathbf{e}^\top \mathbf{W}^{-1}\mathbf{e}\lambda^\top (2\Sigma)^{-1} = \mathbf{e}^\top. \tag{7.11}$$

$$\lambda^\top = \frac{2}{p}\mathbf{e}^\top \boldsymbol{\Sigma}, \tag{7.12}$$

where $p = \mathbf{e}^\top \mathbf{W}^{-1}\mathbf{e}$. Therefore, the allocation matrix is

$$\hat{\mathbf{A}} = \frac{1}{e^\top \mathbf{W}^{-1} e}\mathbf{W}^{-1}\mathbf{e}\mathbf{e}^\top.$$

Proof of Theorem 7.3
It follows from the actuarial fairness condition $\boldsymbol{A}\boldsymbol{\mu} = \boldsymbol{\mu}$ that

$$\boldsymbol{W}^{-1}\boldsymbol{e}\boldsymbol{e}^\top \boldsymbol{\mu} = \boldsymbol{e}^\top \boldsymbol{W}^{-1}\boldsymbol{e}\boldsymbol{\mu},$$

in vector form,

$$\begin{pmatrix} (1/w_1)\sum_{i=1}^n \mu_i \\ (1/w_2)\sum_{i=1}^n \mu_i \\ \vdots \\ (1/w_n)\sum_{i=1}^n \mu_i \end{pmatrix} = \begin{pmatrix} \sum_{i=1}^n (1/w_i)\mu_1 \\ \sum_{i=1}^n (1/w_i)\mu_2 \\ \vdots \\ \sum_{i=1}^n (1/w_i)\mu_n \end{pmatrix}.$$

By solving the linear system in $1/w_i$, we obtain the solution

$$w_i = \frac{1}{\mu_i}.$$

Thus, the allocation matrix is given by

$$\hat{\boldsymbol{A}} = \frac{1}{\boldsymbol{e}^\top \boldsymbol{M}\boldsymbol{e}}\boldsymbol{M}\boldsymbol{e}\boldsymbol{e}^\top.$$

Appendix 7.C Pareto in the Class of Fair Risk Exchanges

Proof Define an augmented objective function $\mathcal{L}$ that includes two additional Lagrange multiplier terms by

$$\mathcal{L} = \text{tr}(\boldsymbol{A}\boldsymbol{\Sigma}\boldsymbol{A}^\top) + (\boldsymbol{e}^\top - \boldsymbol{e}^\top\boldsymbol{A})\boldsymbol{\lambda} + \boldsymbol{\beta}^\top(\boldsymbol{\mu} - \boldsymbol{A}\boldsymbol{\mu}).$$

Utilizing matrix differentiation identities, it follows that

$$0 = \frac{\partial \mathcal{L}}{\partial \boldsymbol{A}} = \boldsymbol{A}\boldsymbol{\Sigma}^\top + \boldsymbol{A}\boldsymbol{\Sigma} - \boldsymbol{e}\boldsymbol{\lambda}^\top - \boldsymbol{\beta}\boldsymbol{\mu}^\top; \tag{7.13}$$

$$0 = \frac{\partial \mathcal{L}}{\partial \boldsymbol{\lambda}} = \boldsymbol{e}^\top - \boldsymbol{e}^\top\boldsymbol{A}; \tag{7.14}$$

$$0 = \frac{\partial \mathcal{L}}{\partial \boldsymbol{\beta}} = \boldsymbol{\mu} - \boldsymbol{A}\boldsymbol{\mu}. \tag{7.15}$$

From (7.13), we can solve for the allocation matrix to find

$$\hat{\boldsymbol{A}} = (\boldsymbol{e}\boldsymbol{\lambda}^\top + \boldsymbol{\beta}\boldsymbol{\mu}^\top)(2\boldsymbol{\Sigma})^{-1}.$$

Inserting this expression into Eqs. (7.14) and (7.15) yields

$$\boldsymbol{e}^\top(\boldsymbol{e}\boldsymbol{\lambda}^\top + \boldsymbol{\beta}\boldsymbol{\mu}^\top)(2\boldsymbol{\Sigma})^{-1} = \boldsymbol{e}^\top. \tag{7.16}$$

$$(\boldsymbol{e}\boldsymbol{\lambda}^\top + \boldsymbol{\beta}\boldsymbol{\mu}^\top)(2\boldsymbol{\Sigma})^{-1}\boldsymbol{\mu} = \boldsymbol{\mu}. \tag{7.17}$$

Solving the constraint Eq. (7.16) for the Lagrange multiplier $\boldsymbol{\lambda}$ gives

$$\boldsymbol{\lambda}^\top = \frac{1}{n}\boldsymbol{e}^\top(2\boldsymbol{\Sigma} - \boldsymbol{\beta}\boldsymbol{\mu}^\top). \tag{7.18}$$

Utilizing Eq. (7.17), we find that

$$\boldsymbol{\beta} = k\boldsymbol{\mu}, \quad \text{where} \quad k = [\boldsymbol{\mu}^\top(2\boldsymbol{\Sigma})^{-1}\boldsymbol{\mu}]^{-1}.$$

Note that (7.18) can be rewritten as

$$\boldsymbol{\lambda}^\top = \frac{1}{n}\boldsymbol{e}^\top(2\boldsymbol{\Sigma} - k\boldsymbol{\mu}\boldsymbol{\mu}^\top).$$

This yields a final expression for the allocation matrix in terms of the agent mean loss vector and covariance matrix given by

$$\hat{\boldsymbol{A}} = \frac{1}{n}\boldsymbol{e}\boldsymbol{e}^\top + k(\boldsymbol{I} - \frac{1}{n}\boldsymbol{e}\boldsymbol{e}^\top)\boldsymbol{\mu}\boldsymbol{\mu}^\top(2\boldsymbol{\Sigma})^{-1},$$

where $\boldsymbol{I}$ is an n-dimensional identity matrix. The above simplifies slightly to

$$\hat{\boldsymbol{A}} = \frac{1}{n}\boldsymbol{e}\boldsymbol{e}^\top + k\left(\boldsymbol{I} - \frac{1}{n}\boldsymbol{e}\boldsymbol{e}^\top\right)\boldsymbol{\mu}\boldsymbol{\mu}^\top\boldsymbol{\Sigma}^{-1} \quad \text{where} \quad k^{-1} = \boldsymbol{\mu}^\top\boldsymbol{\Sigma}^{-1}\boldsymbol{\mu}. \tag{7.19}$$

□

Problems

1. Suppose that a group of participants start with a Pareto optimal risk sharing based on power utility as in Example 6.2. To minimize transaction cost on compensation payments, the group decides to use a least squares risk sharing strategy. Determine the risk exchange between peers i and j in a complete network.
2. Consider the Pareto optimal risk sharing plan in the class of fair risk exchanges. In the two-agent case, show the allocation coefficients α_{11} and α_{22} in (7.6) by solving a quadratic equation.
3. Prove the solution to the altruistic risk sharing plan in Example 7.3 by Karush-Kuhn-Tucker conditions.

Chapter 8
Unified Framework

The purpose of this chapter is to close the gap in product designs in the market practice of decentralized insurance and theoretical models of risk sharing in the literature. While decentralized insurance models often exhibit different forms of business logic, they share some common foundations for risk sharing. We shall reveal some hidden connections in a unified framework, which typically involve a combination of risk sharing and risk transfer techniques. Such a framework enables us to compare and analyze existing business models, and also to draw on the wisdom from business practices and theoretical innovations, and to generate new decentralized insurance schemes.

8.1 Two-Step Process

The decentralized insurance framework can be broken down into several steps. It is often the case to have two steps. The first step is for participants to bring forth their risks and to exchange them with each other within the community in the hope to achieve some risk diversification and mitigation. After the risks are exchanged and distributed back to individuals, some portions of the risks beyond the capacity of participants are then passed on to third parties.

This process can be best explained with analogy to a pizza cooking party as shown in Fig. 8.1. Many cooks bring to the table their favorite ingredients for pizza. They pool their ingredients together to make a pizza. This is in essence similar to the risk exchange where participants put forth their risks and mix their portfolios. Of course, parties may have different preferences of cooking styles and different ways to slice and dice up the pizzas, just as there are also different risk sharing rules. Cooks each eat slices of the pizza to comfortably full and share left-overs with other foodies who enjoy pizza for pleasure. People of varying appetites may be

R. Feng, *Decentralized Insurance*, Springer Actuarial,
https://doi.org/10.1007/978-3-031-29559-1_8

Fig. 8.1 Analogy between pizza cooking party and decentralized insurance

able to eat different amounts. The same can be said about decentralized insurance. Participants accept certain portions of the allocated risks and pass on the rest to third-party professional risk takers for profits. There are also a variety of risk transfer rules by which participants can work with different kinds of risk takers.

It should be pointed out that the two steps may not necessarily be undertaken in a chronological order. There are also business models that may include more than two steps. Even with each step, there could be some mix-and-match of a wide range of rules for different participants. Examples can be found in Sect. 8.4.

8.1.1 Risk Sharing (Mutualization)

A foundation of insurance and risk management is mutualization, which is to pool together risks from individuals and redistribute them among participants. Participants who do not experience losses indirectly pay for claims from those who suffer losses through their premium payments. However, the insurer can make a profit or a loss as the premium collected do not exactly match the claims. Decentralized insurance is to remove or reduce the role of the intermediary. As such, a loss conservation principle is needed to ensure that all losses are absorbed within the community, i.e.

$$S = \sum_{i=1}^{n} X_i = \sum_{i=1}^{n} Y_i.$$

We can formalize the concept of mutualization in general terms.

Table 8.1 Examples of risk sharing rules

Scheme	Symbol	Rule
	Aggregate risk sharing $\boldsymbol{h}^o(\boldsymbol{X})$	
All-to-claimant uniform	$h_i^a(S)$	S/n
Survivor-to-claimant uniform	$h_i^s(S)$	$(1 - I_i)S/(n - N)$, where $I_i = \begin{cases} 1, & i \text{ makes a claim;} \\ 0, & \text{otherwise,} \end{cases}$ and $N = \sum_{i=1}^n I_i$.
Pro rata	$h_i^p(S)$	$(\mathbb{E}[X_i]/\mathbb{E}[S])S$
Quota-share	$h_i^q(S)$	$\alpha_i S + \beta_i$, where $\sum_{i=1}^n \alpha_i = 1, \sum_{i=1}^n \beta_i = 0$.
Conditional mean	$h_i^c(S)$	$\mathbb{E}[X_i \mid S]$
	P2P risk sharing $\boldsymbol{h}(\boldsymbol{X})$	
Minimum variance Pareto optimal		$\hat{\boldsymbol{A}}\boldsymbol{X}$
Least squares		$(1/n)[(\boldsymbol{X} - \boldsymbol{h}^o(\boldsymbol{X})) \otimes \boldsymbol{e}^\top - (\boldsymbol{X} - \boldsymbol{h}^o(\boldsymbol{X}))^\top \otimes \boldsymbol{e}]$

Definition 8.1 A risk sharing rule, $\boldsymbol{h}$, is a mapping that transforms a random vector $\boldsymbol{X} = (X_1, \cdots, X_n)$ into another of the same dimension, $\boldsymbol{h}(\boldsymbol{X}) = (h_1(\boldsymbol{X}), \cdots, h_n(\boldsymbol{X}))$ such that

$$\sum_{i=1}^n h_i(\boldsymbol{X}) = \sum_{i=1}^n X_i.$$

Table 8.1 offers a sample list of risk sharing rules discussed in this paper. However, this list is far from comprehensive, as there are a wide variety of risk sharing rules in the literature. Various properties of risk sharing rules have been studied in Denuit et al. (2022a).

Example 8.1 (*Linear risk sharing*) The risk sharing rule $\boldsymbol{h}$ is called quota-share or linear if there exists a matrix $\boldsymbol{A}$ with column sums being one, i.e. $\boldsymbol{e}^\top \boldsymbol{A} = \boldsymbol{e}^\top$, such that $\boldsymbol{h}(\boldsymbol{X}) = \boldsymbol{A}\boldsymbol{X}$. It is a linear risk sharing rule under the principle of indemnity if all entries of $\boldsymbol{A}$ are non-negative, or in other words, $\boldsymbol{A}$ is a left-stochastic matrix. The principle of indemnity states that an insurance coverage should not provide a compensation that exceeds the insured's loss. If the risks are homogeneous, then the linear risk sharing is actuarially fair when the row sums of $\boldsymbol{A}$ are also one, i.e. $\boldsymbol{A}\boldsymbol{e} = \boldsymbol{e}$. For this reason, it is often desirable to discuss linear risk sharing with doubly stochastic matrix $\boldsymbol{A}$.

8.1.2 *Risk Transfer to Third Party*

Another pillar of insurance and risk management is the transfer of risk from an insured to an insurer. The policyholder pays for a fixed premium in exchange for the insurer to take on the risk beyond his/her own risk-bearing capacity. This mechanism is also present in various forms of decentralized insurance that require ex-ante contributions, such as takaful and P2P insurance. Because the deposits are used to cover losses and any remaining amounts are returned to participants, such a mechanism is in essence the same as asking participants without any deposit to pay ex-post up to the amount of deposit. In other words, the deposit merely serves as a reference point for limited liability for a participant. Any risk beyond the financial capacity of a participant could be transferred to third parties outside the community. Therefore, we can formalize such a process by risk transfer.

Definition 8.2 A risk transfer rule g is a mapping that transforms a positive random variable X into another positive random variable $g(X)$, such that $X - g(X)$ and $g(X)$ are comonotonic.

The risk $g(X)$ is retained by the entity that provides the risk X and the risk $X - g(X)$ is transferred to the third party. In a broader sense, a risk transfer rule can refer to any strategy that involves a contractual shifting from one party to another. The comonotonicity implies that when losses increase, both parties of the risk transfer rule have to bear more severe consequences.

Definition 8.3 Under a given premium principle $\mathcal{P}$, a risk transfer rule g is said to be actuarially fair if $\mathcal{P}[X] = \mathcal{P}[g(X)]$.

It should be pointed out that fairness is defined and discussed in Bühlmann and Jewell (1979) and later studied in general terms in Pazdera et al. (2017).

Example 8.2 *(CAT risk pooling)* In the homogeneous case, the CAT risk transfer scheme is clearly actuarially fair under the expected value premium principle as $\mathbb{E}[g^c(Y)] = \mathbb{E}[Y] = \mathbb{E}[X]$. However, this may not always be true in the heterogeneous case. Recall that the entity j's retained risk is given by (5.8). If we assume $\gamma_j = \alpha_j$, then the actuarial fairness can only be satisfied if $\gamma_j = \alpha_j = \frac{\mathbb{E}[Y_j]}{\mathbb{E}[S]}$. This is because $\mathbb{E}[Y_j] = \gamma_j \mathbb{E}[S]$.

Table 8.2 Examples of risk transfer rules

Scheme	Symbol	Rule
Traditional Insurance	$g^b(Y)$	$\mathbb{E}[Y]$
P2P Insurance	$g^p(Y)$	$\mathbb{E}[(Y-d)_+] + d \wedge Y$
Mutual Aid	$g(Y)$	Y
Takaful Mudarabah	$g^m(Y)$	$d - (1-\rho^m)(d-Y)_+$
Takaful Wakalah	$g^w(Y)$	$d - [(1-\rho^w)d - Y]_+$
Takaful Hybrid	$g^h(Y)$	$d - [d(1-\rho^w) - Y]_+(1-\rho^m)$
CAT Risk Pooling	$g^c(Y)$	$d \wedge Y + \mathbb{E}[(Y \wedge m - d)_+] + [Y-m]_+$
Co-share		βY

8.1.3 Decentralized Insurance Scheme

Definition 8.4 A decentralized insurance scheme with a risk sharing rule $\boldsymbol{h}$ and a risk transfer rule $\boldsymbol{g}$ is a composite mapping that transforms a random vector $\boldsymbol{X}$ into another one of the same dimension, denoted by $\boldsymbol{Y} = \boldsymbol{g} \circ \boldsymbol{h}(\boldsymbol{X})$ where $Y_i = g_i \circ h_i(\boldsymbol{X})$.

It is easy to see that a decentralized insurance scheme of dimension n has a one-to-one correspondence with a risk sharing rule of dimension $n+1$. Under such a risk sharing rule, risks are being exchanged among risk providers represented by $\boldsymbol{X}$ with the arrangement $\boldsymbol{Y}$ in a decentralized network and partially being transferred to a third-party risk taker Z such that

$$\sum_{i=1}^{n} X_i = \sum_{i=1}^{n} Y_i + Z.$$

Such a risk taker can be a traditional insurer, or a takaful operator, or a cohort of investors. In other words, the risk sharing rule of dimension $n+1$ can be represented by $\tilde{\boldsymbol{h}} = (\tilde{h}_1, \cdots, \tilde{h}_{n+1})$ where $\tilde{h}_i = g_i \circ h_i$, for $1 \le i \le n$, and

$$\tilde{h}_{n+1} = \sum_{i=1}^{n} (h_i - g_i \circ h_i).$$

Theorem 8.1 *If both the risk sharing rule* $\boldsymbol{h}$ *and risk transfer rule* $\boldsymbol{g}$ *satisfy the actuarial fairness condition, then the decentralized insurance scheme* $\boldsymbol{Y} = \boldsymbol{g} \circ \boldsymbol{h}(\boldsymbol{X})$ *is actuarially fair, i.e.* $\mathcal{P}[\boldsymbol{Y}] = \mathcal{P}[\boldsymbol{X}]$.

This result follows by definition that $\mathcal{P}[Y_i] = \mathcal{P}[g_i(h_i(\boldsymbol{X}))] = \mathcal{P}[h_i(\boldsymbol{X})] = \mathcal{P}[X_i]$.

While there are other premium principles available, we primarily use the expected value principle $\mathcal{P} = \mathbb{E}$ throughout the chapter for its simplicity and analytical property. For example, we shall show in the next section that, when the risk transfer rules are actuarially fair, many decentralized insurance schemes can be compared by the so-called convex order.

8.2 Ordering of Decentralized Insurance Schemes

We have so far formulated all commonly known forms of risk sharing mechanisms under this overarching decentralized insurance framework. An advantage of such a framework is that these decentralized insurance schemes can be compared quantitatively. The key to a comparative study of various risk sharing mechanisms is the notion of convex order, which we reiterate below.

Definition 8.5 Given two random variables X and Y, if $\mathbb{E}[\phi(X)] \leq \mathbb{E}[\phi(Y)]$ for all convex functions ϕ, provided that the expectations exist, then X is said to be less than Y in the convex order, denoted by $X \leq_{cx} Y$. Given two random vectors $\boldsymbol{X} = (X_1, \cdots, X_n)$ and $\boldsymbol{Y} = (Y_1, \cdots, Y_n)$, if $X_i \leq_{cx} Y_i$ for all $i = 1, \cdots, n$, then $\boldsymbol{X}$ is said to be less than $\boldsymbol{Y}$ componentwise in the convex order, denoted by $\boldsymbol{X} \leq_{ccx} \boldsymbol{Y}$.

8.2.1 *Risk Diversification by Mutualization*

In the general framework, we first show that every participant of a decentralized insurance scheme can benefit from certain types of risk sharing, i.e. $\boldsymbol{h}(\boldsymbol{X}) \leq_{cx} \boldsymbol{X}$. Here we use the aforementioned linear and conditional mean risk sharing rules as examples to illustrate the effect of risk diversification.

Example 8.3 (*Linear risk sharing*) Suppose that the risk vector $\boldsymbol{X}$ is exchangeable. If $\boldsymbol{Y}$ is a linear risk sharing of $\boldsymbol{X}$ with a doubly stochastic matrix $\boldsymbol{A}$, then $\boldsymbol{Y} \leq_{ccx} \boldsymbol{X}$.

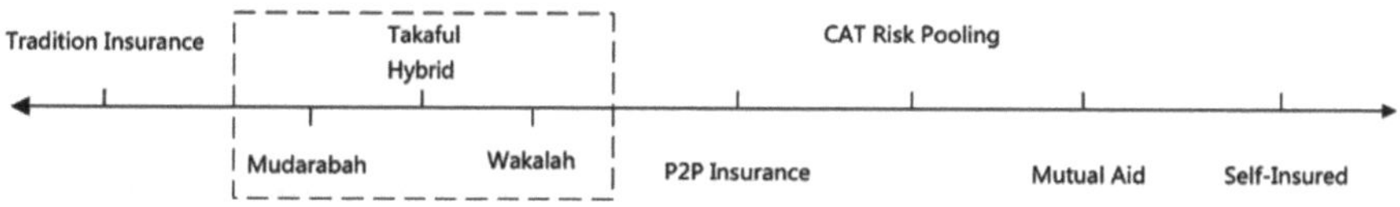

Fig. 8.2 Convex order spectrum of decentralized insurance

This result is known in Charpentier et al. (2021) for the case of independent and identically distributed risks. To make the book self-contained, we give a proof for exchangeable risks in Appendix 8.A.

Example 8.4 (*Conditional mean risk sharing*) If $\boldsymbol{Y}$ is the conditional mean risk sharing of $\boldsymbol{X}$, then $\boldsymbol{Y} \leq_{ccx} \boldsymbol{X}$.

This result is known in Denuit and Dhaene (2012b) and follows immediately from Jensen's inequality for conditional expectation, $\mathbb{E}[\phi(Y_i)] = \mathbb{E}[\phi(\mathbb{E}[X_i|S])] \leq \mathbb{E}[\mathbb{E}[\phi(X_i)|S]] = \mathbb{E}[\phi(X_i)]$.

8.2.2 *Risk Reduction by Transfer*

After the risk exchange among peers, it is expected that every participant reduces further his/her exposure by some risk transfer rule, i.e. $\boldsymbol{g}(\boldsymbol{h}(\boldsymbol{X})) \leq_{ccx} \boldsymbol{h}(\boldsymbol{X})$. Here, we turn our attention to the comparison of risk transfer rules. It turns out that all risk transfer rules can be put in an increasing convex order from the left to right on a spectrum in Fig. 8.2.

In the comparison, we assume that all risk transfer rules are actuarially fair and all profit-sharing ratios are between 0 and 1. In particular, we use the expected value principle as the premium principle under consideration. In other words, each participant is only willing to undertake a risk in exchange for a fixed premium of an amount equal to the expectation of the accepted risk. In the context of convex ordering, we assume that no friction cost, such as commission, expense, or underwriting profit, is allowed for undertaking risks.

Theorem 8.2 (Risk Reduction by Transfer) *In the absence of any friction cost, the following risk sharing schemes increase in convex order,*

$$\mathbb{E}[X_i] \leq_{cx} g^w(Y_i) \leq_{cx} g^p(Y_i) \leq_{cx} g^c(Y_i) \leq_{cx} Y_i.$$

From this convex order, we can also derive that $\mathbb{V}(Y_i) \geq \mathbb{V}(g^c(Y_i)) \geq \mathbb{V}(g^p(Y_i)) \geq \mathbb{V}(g^w(Y_i)) \geq 0 = \mathbb{V}(\mathbb{E}[X_i])$. On one end of the spectrum, traditional insurance policyholders transfer all risks (assume there is no deductible and coinsurance, etc.) to the insurer and thus take the least amount of risk. On the other end of the spectrum, the entities take all risks on their own and thus have the largest amount of risk. In theory, one can choose a decentralized insurance scheme according to his/her risk preference and adjust its risk level by tuning parameters. The proof is provided in Sect. 8.4.

Note that a different risk transfer rule may be applied to each exchanged risk. In other words, $Y_i = h_i(\boldsymbol{X})$ may be different for each i. However, the discussion in this section is only restricted to risk transfer, which is considered as a separate procedure from risk sharing. The comparison of risk transfer rules can be independent of risk sharing rules. Therefore, we shall suppress the subscript i in Y_i for notation brevity. For the ease of exposition, we shall add model parameters in the notation of various risk sharing rules.

Theorem 8.3 (P2P Insurance) *Consider two P2P insurance contracts that have different deposit amounts. If $d < d'$, then $g^p(Y; d) \leq_{cx} g^p(Y; d')$.*

The higher the deposit the more a participant takes own risk. The result confirms that the retained risk from entering a contract with a high deposit is less risky in convex order than that from a contract with a low deposit. The level of deposit d can be used as a parameter to adjust the level of decentralization.

Theorem 8.4 (Catastrophe Risk Pooling) *Consider two CAT risk pooling plans with different attachment points or exhaustion points.*

(i) If the attachment point d is the same and the exhaustion points are such that $m < m'$, then $g^c(Y; m') \leq_{cx} g^c(Y; m)$.
(ii) If the exhaustion point m is the same and the attachment points are such that $d < d'$, then $g^c(Y; d) \leq_{cx} g^c(Y; d')$.

In CAT risk pooling, the range $[d, m]$ indicates the amount of risk to be ceded to a third party. The more risk ceded, the less risky the retained risk. The first two parts of the theorem confirm that in terms of convex order. The change in the range of attachment to exhaustion points is also reflected in the premium of coverage. Parts (i) and (ii) indicate that, when either of the attachment point or the exhaustion point is the same, the plan with a high premium is less risky than that with a low premium, or put in a different way, the retained risk decreases in convex order with the increase of the premium for the ceded risk.

Theorem 8.5 (Takaful) *The takaful models decrease in convex order in the order of wakalah, hybrid, and mudarabah, i.e. $g^m(Y) \leq_{cx} g^h(Y) \leq_{cx} g^w(Y)$.*

While all three models enable participants to transfer away excessive losses beyond their initial contribution levels, they change the distribution of losses below the contribution level for participants in different ways. In the wakalah business model, participants pay a fixed wakalah fee in addition to shared cost below the contribution level in a manner similar to that of P2P insurance. In the mudarabah model, participants give up a portion of the surplus (cashback for losses below the contribution level). While both remove the upper-tail uncertainty (large losses), the latter increases the sizes of lower-tail losses (small losses) for participants compared with the former, hence making the loss distribution under the mudarabah model more concentrated than that of the wakalah model. Hence, in general, the mudarabah model results in less uncertainty for participants than the wakalah model. The hybrid model requires participants to pay a fixed upfront fee and then to share a portion of the surplus with the operator. As it accommodates features of both mudarabah and wakalah models, the mudarabah and wakalah models can be viewed as extreme cases where $\rho^{hw} = 0$ and $\rho^{hm} = 0$, respectively. It is not surprising that the hybrid model sits in between two extreme cases in terms of convex order.

Theorem 8.6 (Takaful) *Consider takaful business models with different contribution levels. If $d < d'$, then $g^w(Y; d) \leq_{cx} g^w(Y; d')$ and $g^m(Y; d) \leq_{cx} g^m(Y; d')$. If $d < d'$ and $(1 - \rho^{hw'})d' = (1 - \rho^{hw})d$, then $g^h(Y; d) \leq_{cx} g^h(Y; d')$.*

Intuitively speaking, for any form of takaful, the higher the contribution level, the more risk the participant effectively takes on. Therefore, a takaful contract with a high contribution tends to be more risky than one with a low contribution, as more cash is exposed to the underlying risk. Note that the hybrid model has two parameters including the profit-sharing ratio, ρ^{hm}, and the wakalah fee rate ρ^{hw}. Even with different contribution levels, we could neutralize the impact of one parameter. Therefore, we assume that $(1 - \rho^{hw'})d' = (1 - \rho^{hw})d$. Note that $(1 - \rho^{hw})d$ is the post-fee contribution level above which all losses can be covered by a qard hassan loan from the operator. When this level is fixed, the contracts have the same liability for the operator. Then the contract with a high contribution level comes with higher uncertainty than one with a low contribution level.

8.3 Decentralized Insurance for Heterogeneous Risks

We have so far mostly introduced various business models in Chap. 5 for homogeneous risks. For the ease of presentation, we discuss the extension of these models for heterogeneous risks in the above-stated framework and state the conditions under which actuarial fairness is guaranteed.

8.3.1 P2P Insurance

(Mutualization). At the core of such an arrangement is the underlying risk sharing mechanism. Peer i brings the risk X_i into the risk sharing. The losses are exchanged among all participants for any actuarially fair comonotonic risk sharing rule $\boldsymbol{h}(S)$. Thus, peer i is responsible for the shared cost $h_i(S)$. Recall the loss conservation condition that $S = \sum_{i=1}^n X_i = \sum_{i=1}^n h_i(S)$.

(Risk transfer to an insurer). Each participant names his/her own deposit into a funding pool. The P2P insurance model limits the liability of each participant up to the initial deposit. Let d_i be the initial deposit from peer i, which measures the maximum level of uncertainty that the participant is willing to take on. The participant retains the risk $h_i(S) \wedge d_i$ and transfers the risk $(h_i(S) - d_i)_+$ to an insurer. Participant i agrees to purchase an excess-of-loss insurance policy from an insurer, say $\pi_i = \mathbb{E}[(h_i(S) - d_i)_+]$. Thus, the participant's actual liability is given by

$$g^P[h_i(S)] = d_i \wedge h_i(S) + \pi_i.$$

Note that the choice of d_i does not matter throughout the previous analysis. It follows from Theorem 8.1 that, as long as the risk sharing rule is actuarially fair under the expected value principle, the P2P insurance is always actuarially fair. Contrary to the current practice where deposits/initial premiums are often specified, one could in theory have the complete freedom to choose his or her own deposit while keeping the mechanism fair for all.

Example 8.5 (*P2P insurance with conditional mean risk sharing*) Suppose that there are two participants in the P2P insurance whose losses are independent and normally distributed with mean and standard deviation in the table below.

Peer	μ_i	σ_i	Scenario #1	Scenario #2
#1	4	2	7	10
#2	6	4	3	20

First, consider the conditional mean risk sharing rule for normal random variables,

$$h_i(s) = \mu_i + \frac{\sigma_i^2}{\sigma_1^2 + \sigma_2^2}(s - (\mu_1 + \mu_2)), \qquad i = 1, 2.$$

This rule is derived and discussed earlier with details in Example 6.9. Specifically for this application, we must have

$$h_1(s) = 4 + \frac{4}{4+16}(s - 10) = \frac{s}{5} + 2;$$
$$h_2(s) = 6 + \frac{16}{4+16}(s - 10) = \frac{4s}{5} - 2.$$

Observe that the actual aggregate loss $S = X_1 + X_2$ is also normally distributed with mean 10 and variance 20. The shared risk for peer #1, $Y_1 = h_1(S)$, has mean $\mathbb{E}[Y_1] = 10/5 + 2 = 4 = \mathbb{E}[X_1]$ and variance

$$\mathbb{V}[Y_1] = 20/25 = 4/5 < \mathbb{V}[X_1] = 4.$$

The shared risk for peer #2, $Y_2 = h_2(S)$, has mean $\mathbb{E}[Y_2] = 4 \times 10/5 - 2 = 6 = \mathbb{E}[X_2]$ and variance

$$\mathbb{V}[Y_2] = 16 \times 20/25 = 64/5 < \mathbb{V}[X_2] = 16.$$

This risk sharing rule implies that, even if there is no loss from either participant, peer #1 has to make a payment of $h_1(0) = 2$ to peer #2.

None of the two participants wants to take on unlimited losses. They agree to construct a P2P insurance scheme, in which each makes a deposit up to his/her risk-bearing capacity. Suppose that the total initial deposit from both participants is given by $a = 15$, of which peer #1 contributes $a_1 = h_1(a) = 5$ and peer #2 contributes $a_2 = h_2(a) = 10$. In addition, they are willing to split the cost of reinsurance. Recall from Exercise 2.3 that

$$\mathbb{E}[(X-a)_+] = \mu + \sigma \frac{\phi\left(\frac{a-\mu}{\sigma}\right)}{1-\Phi\left(\frac{a-\mu}{\sigma}\right)} - a\left(1 - \Phi\left(\frac{a-\mu}{\sigma}\right)\right).$$

Thus, the total cost of reinsurance is given by

$$\pi = \mathbb{E}[(S-a)_+] = 10 + \sqrt{20}\frac{\phi\left(\frac{15-10}{\sqrt{20}}\right)}{1-\Phi\left(\frac{15-10}{\sqrt{20}}\right)} - 15\left(1-\Phi\left(\frac{15-10}{\sqrt{20}}\right)\right) \simeq 15.3.$$

The two peers share the cost by

$$\pi_1 = \mathbb{E}[(Y_1 - a_1)_+] = 4 + \sqrt{\frac{4}{5}}\frac{\phi\left(\frac{5-4}{\sqrt{\frac{4}{5}}}\right)}{1-\Phi\left(\frac{5-4}{\sqrt{\frac{4}{5}}}\right)} - 5\left(1-\Phi\left(\frac{5-4}{\sqrt{\frac{4}{5}}}\right)\right) \simeq 4.8;$$

$$\pi_2 = \mathbb{E}[(Y_2 - a_2)_+] = 6 + \sqrt{\frac{16}{5}}\frac{\phi\left(\frac{10-6}{\sqrt{\frac{16}{5}}}\right)}{1-\Phi\left(\frac{10-6}{\sqrt{\frac{16}{5}}}\right)} - 10\left(1-\Phi\left(\frac{10-6}{\sqrt{\frac{16}{5}}}\right)\right) \simeq 10.5.$$

It is clear that $\pi = \pi_1 + \pi_2$. This result is not unexpected because

$$(S-a)_+ = (h_1(S) - h_1(a))_+ + (h_2(S) - h_2(a))_+ \tag{8.1}$$

for increasing functions h_1 and h_2. When $S \geq a$, we must have $h_1(S) \geq h_1(a)$ and $h_2(S) \geq h_2(a)$. In view of the fact that $h_1(s) + h_2(s) = s$, we obtain

$$(S-a)_+ = S - a = (h_1(S) - h_1(a)) + (h_2(S) - h_2(a)) = (h_1(S) - h_1(a))_+ + (h_2(S) - h_2(a))_+.$$

When $S < a$, both sides of (8.1) are zero. Hence, the proof of (8.1) is complete. It follows immediately that $\pi = \pi_1 + \pi_2$.

If there is any unspent amount from the deposit, $(a - S)_+$, then peer #1 receives a cashback of the amount $(a_1 - Y_1)_+$ and peer #2 receives a cashback of the amount $(a_2 - Y_2)_+$. As shown earlier, these should be the exact amount to be split between two participants, i.e. $(a - S)_+ = (a_1 - Y_1)_+ + (a_2 - Y_2)_+$.

Consider two scenarios of actual losses from the two participants in the table. In Scenario #1, the total loss is given by $S = 10$. According to the P2P insurance scheme established earlier, the community covers all losses within its capacity, $S \wedge a = 10$, of which the amount $h_1(10) = 10/5 + 2 = 4$ is deducted from peer #1's account and the amount $h_2(10) = 4 \times 10/5 - 2 = 6$ is deducted from peer #2's account. There is an unspent amount $(a - S)_+ = 15 - 10 = 5$, of which $(a_1 - h_1(10))_+ = 5 - 4 = 1$ is returned to peer #1 and $(a_2 - h_2(10))_+ = 10 - 6 = 4$ goes back to peer #2.

In Scenario #2, the total loss $S = 30$ is partially paid by the community up to its risk-bearing capacity, i.e. $S \wedge a = 15$. The rest is covered by the reinsurer. Of the amount paid by the community, peer #1 carries the amount $h_1(S) \wedge a_1 = (30/5 + 2) \wedge 5 = 8 \wedge 5 = 5$ and peer #2 carries the amount $h_2(S) \wedge a_2 = (4 \times 30/5 - 2) \wedge 10 = 22 \wedge 10 = 10$. Since both participants have reached their risk-bearing capacities, there is no cashback from their initial deposits.

One can also verify the actuarial fairness in this scheme. Note that the post-exchange risk for peer i can be represented by

$$g^p(Y_i) = a_i + \pi_i - (a_i - Y_i)_+,$$

where the two terms are the deposit and the reinsurance premium and the last term is the cashback. Observe that

$$\begin{aligned}\mathbb{E}[g^p(Y_i)] &= \mathbb{E}[a_i + \mathbb{E}[(a_i - Y_i)_-] - (a_i - Y_i)_+] \\ &= \mathbb{E}[a_i - (a_i - Y_i)] = \mathbb{E}[Y_i] = \mathbb{E}[X_i].\end{aligned}$$

If the two participants purchase their own excess-of-loss policies, the exact value of attachment point a_i does not actually matter. One could choose attachment points other than $h_i(a)$. The actuarial fairness holds regardless of the choice of a_i.

8.3.2 Takaful

Suppose that each takaful participant makes a contribution to the takaful fund, say d_i for participant i. For simplicity, we assume that the takaful fund retains zero balance and sets the surplus distribution threshold at zero. For better comparison, we only consider a single-period model for takaful in this b, which closely resembles that of the P2P insurance.

Mudarabah Model

In the unified framework, the first step for the mudarabah model is to exchange risks among all participants so that participant i is essentially liable to the shared cost $h_i(\boldsymbol{X})$ for some actuarially fair risk sharing rule. The participant retains the risk $h_i(\boldsymbol{X}) \wedge d_i$ and cedes a portion of his/her risk, $(g^m[h_i(\boldsymbol{X})] - d_i)_+$ to the takaful operator in the form of a qard hassan loan. Takaful participants share with the takaful operator a portion of the potential surplus. Let ρ_i^m be the profit-sharing ratio for participant i. The net cost of participation for the participant is the allocated cost less the ceded risk plus the added cost of profit-sharing, i.e.

$$g^m[h_i(\boldsymbol{X})] = h_i(\boldsymbol{X}) \wedge d_i + \rho_i^m(d_i - h_i(\boldsymbol{X}) \wedge d_i) = d_i - (1 - \rho_i^m)[d_i - h_i(\boldsymbol{X})]_+.$$

The first expression presents a breakdown of the cost into two components, namely the participant's liability up to his/her initial deposit and a percentage of the surplus to be shared with the takaful operator. The second expression offers an alternative interpretation in terms of the participant's incoming and outgoing cash flows that the net cost is the initial deposit less the cashback which is a percentage of the surplus.

The operator commits to a qard hassan loan in the time of deficit for the takaful fund in exchange for the profit-sharing in the time of surplus. In return, participant i gives up the profit $\rho_i^m(d_i - Y_i)_+$ to the operator. By the actuarial fairness condition, it is clear that the profit-sharing ratio has to satisfy

$$\mathbb{E}[\left(g^m[h_i(\boldsymbol{X})] - d_i\right)_+] = \rho_i^m \mathbb{E}[(d_i - g^m[h_i(\boldsymbol{X})])_+].$$

Note that the assumption $0 < \rho_i^m < 1$ requires that $\mathbb{E}[(d_i - g^m[h_i(\boldsymbol{X})])_-] < \mathbb{E}[(d_i - g^m[h_i(\boldsymbol{X})])_+]$, which means that the deficit has a smaller mean than the surplus. It should be pointed out that in the practice of takaful it is uncommon to set individualized profit-sharing ratios. But, if the underlying risk sharing rule is pro rata, i.e. $h_i(\boldsymbol{X}) = (d_i/d)S$, then the profit-sharing ratio is the same for all participants, i.e. $\rho_1^m = \cdots = \rho_n^m = \rho^m$ where

$$\rho^m = \frac{\mathbb{E}[(S/d - 1)_+]}{\mathbb{E}[(1 - S/d)_+]}.$$

Wakalah Model

The distinct feature of the wakalah model is that the operator's fee is charged to the deposit before claims are made. It can be considered as a special case of P2P insurance with an actuarially fair risk transfer premium

$$\pi_i = \rho_i^w d_i = \mathbb{E}[(h_i(\boldsymbol{X}) - (1-\rho_i^w)d_i)_+].$$

The above equation is true because the left side $\pi_i = \rho_i^w d_i$ is the premium collected by the takaful operator (insurer) and the right side is the expected loss for the operator (insurer). As the scheme is actuarially fair for all participants, the insurer has to charge the premium in such a way that the expected income (premium) must match the expected outgo (claim). In the heterogeneous case, let ρ_i^w be the operator's fee rate as a percentage of the deposit for participant i. The participant retains the risk $h_i(\boldsymbol{X}) \wedge (1-\rho_i^w)d_i$ and transfers the risk $[h_i(\boldsymbol{X}) - (1-\rho_i^w)d_i]_+$ to the operator, who provides a qard hassan loan when the loss is large. In exchange, the operator charges $\rho_i^w d_i$ as the wakalah fee. Therefore, the net cost for participant i is given by

$$g^w[h_i(\boldsymbol{X})] = h_i(\boldsymbol{X}) \wedge (1-\rho_i^w)d_i + \rho_i^w d_i = d_i - [(1-\rho_i^w)d_i - h_i(\boldsymbol{X})]_+.$$

In a similar manner, there are two ways to interpret the net cost of participation. The first expression is made up of the participant's liability up to post-fee deposit and the wakalah fee. The second expression shows the participant's incoming and outgoing cash flows. The participant pays the initial deposit less the cashback with the arrangement of an operator in the wakalah model and pays the actual shared cost $h_i(\boldsymbol{X})$ without the operator. Suppose that $h_i(\boldsymbol{X})$ is non-negative for all $i = 1, \ldots, n$. In order to guarantee actuarial fairness, it must hold that $d_i \geq \mathbb{E}[Y_i]$ and that

$$\rho_i^w = \frac{\mathbb{E}[(Y_i - (1-\rho_i^w)d_i)_+]}{d_i}.$$

Hybrid model

The hybrid model incorporates both types of operator's fees, including the profit-sharing ratio, ρ_i^{hm}, and the wakalah fee rate, ρ_i^{hw}. Like in the wakalah model, a fixed fee $\rho_i^{hw} d_i$ is deducted from the participant i's account. Furthermore, like in the mudarabah model, participant i retains the risk $h_i(\boldsymbol{X}) \wedge (1-\rho_i^{hw})d_i$ and transfers the risk $[h_i(\boldsymbol{X}) - (1-\rho_i^{hw})d_i]_+$ to the operator. Therefore, the cost for participant i in the hybrid model is given by

$$g^h(h_i(\boldsymbol{X})) = d_i - [d_i(1-\rho_i^{hw}) - h_i(\boldsymbol{X})]_+(1-\rho_i^{hm}).$$

The hybrid model can be considered as the Wakalah-Mudarabah composite. In view of (5.5) and (5.6), we observe that

$$\begin{aligned}g^m(g^w(h_i(\boldsymbol{X}))) &= (1-\rho_i^{hm})[(1-\rho_i^{hw})d_i \wedge h_i(\boldsymbol{X}) + \rho_i^{hw} d_i] + \rho_i^{hm} d_i \\ &= (1-\rho_i^{hm})[(1-\rho_i^{hw})d_i \wedge h_i(\boldsymbol{X})] + (\rho_i^{hw} + \rho_i^{hm} - \rho_i^{hw}\rho_i^{hm})d_i \\ &= d_i - [d_i(1-\rho_i^{hw}) - h_i(\boldsymbol{X})]_+(1-\rho_i^{hm}) = g^h(h_i(\boldsymbol{X})).\end{aligned}$$

To ensure actuarial fairness, the fee rates ρ_i^{hm} and ρ_i^{hw} must satisfy the equation

$$(1-\rho_i^{hm})\mathbb{E}[((1-\rho_i^{hw})d_i - h_i(\boldsymbol{X}))_+] = d_i - \mathbb{E}(h_i(\boldsymbol{X})).$$

Since the left-hand side of the equation is non-negative, then so is the right-hand side. Thus, $d_i \geq \mathbb{E}(h_i(\boldsymbol{X}))$. Therefore, under the fairness assumption, we have

$$\rho_i^{hw} = \frac{\mathbb{E}[(h_i(\boldsymbol{X}) - (1-\rho_i^{hw})d_i)_+] - \rho_i^{hm}\mathbb{E}[((1-\rho_i^{hw})d_i - h_i(\boldsymbol{X}))_+]}{d_i}.$$

8.4 Composite Decentralized Insurance Schemes

We have so far discussed various existing forms of decentralized insurance. The unified framework enables us to propose new forms with greater flexibility. The first three examples are compositions of risk transfer rules and the last offers a composition of risk transfer and risk sharing rules.

8.4.1 Mudarabah-Wakalah Composite

In Sect. 8.3, we learned that the hybrid takaful model is a composition of the wakalah model and mudarabah model. It is also possible to switch the order of the composition and construct a new form by

$$g^w \circ g^m(Y) = g^w[g^m(Y)] = d - [(1-\rho^m)(d-Y)_+ - \rho^w d]_+.$$

In such a model, participants are first asked to share a percentage, ρ^m, of any surplus with the operator. In addition, participants also pay a fixed fee $\rho^w d$ out of their profits. In other words, only when there is a profit, a participant is asked to pay both a fixed fee and a variable fee. The actuarial fairness condition $\mathbb{E}[g^w \circ g^m(Y)] = \mathbb{E}[Y]$ implies that

$$d - \mathbb{E}(Y) = \mathbb{E}\left[\left((1-\rho^m)(d-Y)_+ - \rho^w d\right)_+\right].$$

Such a model transfers a greater level of uncertainty to the operator. This is because the operator can only receive a compensation when the takaful fund yields a surplus and the operator can only receive the wakalah fee when the participants' surplus is sufficient to cover the fixed fee. Therefore, it is less likely for the operator to receive an equal amount of compensation as other models. Because $g^w(X) \leq_{cx} X$, it is easy to see that $g^w[g^m(Y)] \leq_{cx} g^m(Y)$. In view of Theorem 8.5, we observe that the mudarabah-wakalah composite is smaller in convex order than existing takaful models.

$$g^w \circ g^m(Y) \leq_{cx} g^m(Y) \leq_{cx} g^h(Y) \leq_{cx} g^w(Y).$$

8.4.2 CAT-Mudarabah Composite

We can also devise a new decentralized insurance scheme by combining the takaful mudarabah model and the CAT risk pooling. An arrangement is first made with the CAT risk pooling where participants transfer part of their risks, say with an attachment point d and an exhaustion point m such that $m > d$, to third parties at the cost of an insurance premium $k(Y) = \mathbb{E}[(Y \wedge m - d)_+]$. Therefore, the cost of participation is given by $g^c(Y) = d \wedge Y + k(Y) + [Y - m]_+$. To avoid excessive losses beyond the exhaustion, participants then enter a second arrangement in the form of the takaful mudarabah model. Let ρ^{mc} be the profit-sharing ratio of the mudarabah model and d^m be the initial deposit for the takaful fund. The composite risk transfer rule is then given by

$$(g^m \circ g^c)(Y) = d^m - (1 - \rho^{mc}) \left[d^m - (d \wedge Y + k(Y)) - (Y - m)_+\right]_+ .$$

To ensure actuarial fairness, we require that $\rho^{mc} = \frac{\mathbb{E}[(d^m - g^c(Y))_-]}{\mathbb{E}[(d^m - g^c(Y)_+)]}$. Alternatively, the profit-sharing ratio can be determined by

$$(1 - \rho^{mc})\mathbb{E}\left[d^m - (d \wedge Y + k(Y)) - (Y - m)_+\right]_+ = d^m - \mathbb{E}(Y).$$

It has been proven in Theorem 8.2 that $g^m(Y) \leq_{cx} Y$. Compared with the traditional CAT risk pooling, we obtain

$$(g^m \circ g^c)(Y) \leq_{cx} g^c(Y).$$

In other words, on the spectrum of convex order, the CAT-Mudarabah composite scheme lies in between the Mudarabah scheme on the left and the CAT risk sharing on the right.

8.4.3 Cantor Risk Sharing

The construction of CAT risk pooling scheme has some resemblance to that of a Cantor set studied in real analysis. Recall that a cantor set is obtained by iteratively

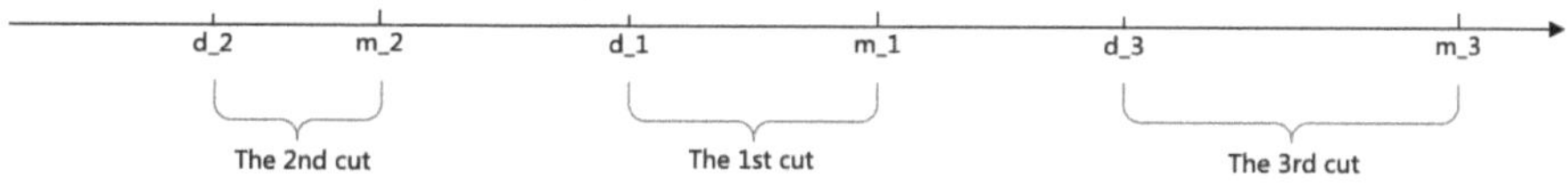

Fig. 8.3 Cantor risk sharing

deleting an open middle third from a set of line segments. We can use a similar idea to develop a risk pooling scheme that transfers a set of different ranges of risk levels to third parties.

Recall that the essence of the CAT risk pooling is to transfer a portion of the underlying risk to a third party. Suppose that the ceded risk lies in the interval (d_i, m_i) for some $0 < d_i < m_i < \infty$. Then the retained risk can be represented by the risk transfer scheme

$$g^{(i)}(Y) := d_i \wedge Y + (Y - m_i)_+.$$

It is possible that one can transfer risks to third parties in a way similar to how a Cantor set is constructed. Suppose that an agent first separates its liability into three layers as shown in Fig. 8.3. The top layer $[m_1, \infty)$, the middle layer (d_1, m_1), and the bottom layer $(-\infty, d_1]$. The agent first transfers the middle layer of its risk to a third party who takes modest risks. The risk transfer scheme can be represented by

$$g^{(1)}(Y) := d_1 \wedge Y + (Y - m_1)_+.$$

The agent may further decide to transfer a middle section (d_2, m_2) of the bottom layer to another third party who takes on small risks and retains the risk

$$\begin{aligned} g^{(2)}(g^{(1)}(Y)) &= d_2 \wedge [d_1 \wedge Y + (Y - m_1)_+] + (d_1 \wedge Y + (Y - m_1)_+ - m_2)_+ \\ &= d_2 \wedge Y + (Y - m_2)_+ \wedge (d_1 - m_2) + (Y - m_1)_+. \end{aligned}$$

Similarly, the agent can transfer a middle section (d_3, m_3) of the large risk to another third party with the capacity for rare and large losses. Then the agent retains the risk

$$\begin{aligned} &g^{(3)}(g^{(2)}(g^{(1)}(Y))) = d_3 \wedge [g^{(2)}(g^{(1)}(Y))] + (g^{(2)}(g^{(1)}(Y)) - m_3)_+ \\ &= d_2 \wedge Y + (Y - m_2)_+ \wedge (d_1 - m_2) + (Y - m_1)_+ \wedge (d_3 - m_1) + (Y - m_3)_+. \end{aligned}$$

The procedure can be carried out iteratively. Set $\mathbb{C}_n = \cup_{i=1}^n (d_i, m_i)$ to be the union of all non-overlapping intervals (cuts). Then the agent ends up with the risk with all those layers removed,

$$\lim_{n\to\infty} g^{(1)} \circ g^{(2)} \cdots \circ g^{(n)}(Y).$$

To ensure actuarial fairness, we need to add back the risk premium $\pi^n = \sum_{i=1}^n \mathbb{E}[(Y - d_i)_+ \wedge (m_i - d_i)]$. Then the fair Cantor risk sharing becomes

$$g^n(Y) = g^{(1)} \circ g^{(2)} \cdots \circ g^{(n)}(Y) + \pi^n.$$

One can show that fair Cantor risk sharing models decrease in convex order with the number of ceded layers, i.e. $g^n(Y) \leq_{cx} g^{n-1}(Y)$. It is intuitive that the more risks are ceded to third parties, the less uncertainty is retained with the ceding party. The fair Cantor risk sharing provides a tool for risk reduction.

8.4.4 General CAT Risk Pooling

In this section, we shall reformulate and generalize the practice of catastrophe risk pooling outlined in Fig. 5.8. Suppose that the actual risks of participating entities are denoted by $\tilde{\boldsymbol{X}} = (\tilde{X}_1, \cdots, \tilde{X}_n)$. As discussed in Sect. 5.5, the cat risk pooling process can be broken down into four steps: (1) risk transfer from entities to a pool, (2) risk aggregation of ceded risks in the pool, (3) risk transfer from the pool to a (re)-insurer, and (4) risk allocation of retained aggregate risk to entities. Such a process can be generalized and summarized in three steps:

- *Risk transfer from entities to a pool.* In the first step, the participating entities decide on risks to be retained, denoted by $\boldsymbol{g}^{(1)}(\tilde{\boldsymbol{X}})$. The rest is ceded to the pool, i.e.

$$\boldsymbol{X} = \tilde{\boldsymbol{X}} - \boldsymbol{g}^{(1)}(\tilde{\boldsymbol{X}}),$$

  where $\boldsymbol{g}^{(1)}(\tilde{\boldsymbol{X}}) = (g_1^{(1)}(\tilde{X}_1), \cdots, g_n^{(1)}(\tilde{X}_n))^\top$ and $g_j^{(1)}$'s are risk transfer rules. In (5.7), the risk transfer rule is given by $g_j^{(1)}(y) = y - \beta_j(h_j \wedge (y - a_j)_+)$.
- *Risk sharing among the entities in the pool.* The post-exchange risks are determined by

$$\boldsymbol{Y} = \boldsymbol{h}(\boldsymbol{X}),$$

  for some general risk sharing rule $\boldsymbol{h}$. In correspondence with (5.8), the risk sharing rule is given by $h_j(\boldsymbol{X}) = \gamma_j \boldsymbol{e}^\top \boldsymbol{X}$.
- *Risk transfer from the pool to a (re)insurer.* The entities then accept risks in accordance with their own risk tolerance and transfer the rest to a third party. Therefore, the entities carry two types of risks, namely the self-retained risk and the allocated risk from the pool,

$$\tilde{\boldsymbol{Y}} = \boldsymbol{g}^{(1)}(\tilde{\boldsymbol{X}}) + \boldsymbol{g}^{(2)}(\boldsymbol{Y}),$$

  for some risk transfer rule $\boldsymbol{g}^{(2)}$. In correspondence with (5.8), the risk transfer rule is given by $g_j^{(2)}(Y_j) = Y_j \wedge A_j + (Y_j - (A_j + H_j))_+ + \alpha_j c$ where $A_j = \gamma_j A$ and $H_j = \gamma_j H$. Even though this is not explicitly stated in the CAT arrangement, participants effectively split the aggregate attachment point A and coverage limit H into individual parts $\{A_1, \cdots, A_n\}$, $\{H_1, \cdots, H_n\}$, and then transfer their own shares of aggregate risk to insurers in accordance with these individual attachment points and coverage limits.

Note that the aggregation-transfer-allocation process used in practice is in fact a special case of the sharing-transfer composition. As discussed in the case of CAT risk pool in Sect. 5.5, this decentralized insurance scheme is actuarially fair when $\gamma_j = \alpha_j = \mathbb{E}[X_j]/\mathbb{E}[S]$, where $S = \boldsymbol{e}^\top \boldsymbol{X}$. In summary, the general CAT risk pooling can be represented as

$$\tilde{\boldsymbol{Y}} = \boldsymbol{g}^{(1)}\left(\tilde{\boldsymbol{X}}\right) + \boldsymbol{g}^{(2)}\left(\boldsymbol{h}\left(\tilde{\boldsymbol{X}} - \boldsymbol{g}^{(1)}\left(\tilde{\boldsymbol{X}}\right)\right)\right).$$

In each of the three steps, one can use any of the aforementioned risk sharing or risk transfer rules or others in the literature to develop new schemes to meet certain desirable standards or criteria.

8.5 Case Study: Xianghubao

In this section, we provide a simple case to show the framework can be used to analyze the current market practice of online mutual aid. Xianghubao used to be the largest mutual aid platform in China, run by Ant Financial, a subsidiary of e-commerce giant Alibaba. Xianghubao was introduced in October 2018 and in less than a year amassed more than 100 million users.

When a member joins the platform, she signs a contract, known as the mutual aid community agreement, which spells out the terms of financial arrangement. According to the agreement, she is obligated to split the cost of mutual aid to others at the end of each month. There is a waiting period of 90 days before she is eligible for claiming mutual aid upon diagnosis of critical illness. The benefit amount is specified in Table 8.3, which varies by age group and the severity of covered illness.

All medical terms of critical illness under the coverage are named in Articles 7.1 and 7.2 of the agreement. When an eligible member is diagnosed with a covered illness, she can make a claim for mutual aid. Her case is subject to reviews by medical professionals and a jury panel consisting of member representatives. If approved, she is expected to receive the specified benefit amount. After all claims are determined and aggregated at the end of each month, the platform determines the cost to be carried by each member using the following formula:

Table 8.3 *Xianghubao* mutual aid amount

Age group	Mild critical illness	Severe critical illness
30 days to 39-year-old	50,000 yuan	300,000 yuan
40- to 59-year-old	50,000 yuan	100,000 yuan

$$\begin{aligned}&\text{per member per month cost}\\&=\frac{\text{total benefit}+\text{management fee}-\text{remainder}}{\text{total number of participants on the platform}}.\end{aligned}$$

Total benefit is the sum of all benefit payments to claimants. Management fee is 8% of all benefit payments. The remainder is used to make sure that the per member cost is rounded up to the nearest fen, which is the smallest denomination in Chinese currency.

The Xianghubao mutual aid plan is in essence a variation of quota-share plan discussed earlier. Let us first reiterate a quota-share plan and then make the connection to mutual aid. Consider a heterogeneous pool of n participants that can be broken down into k classes of homogeneous risks. Let us denote by n_k the size of the kth risk class. Suppose that we do not distinguish allocations by who receives benefits and only allocate the costs according to risk class. Therefore, we shall use α_{ij} for the generic allocation rate between an arbitrary member in the ith risk class and an arbitrary member in the jth risk class for $i, j = 1, \cdots, k$. Let s_j be the benefit amount for a claimant in the jth risk class. Thus, the net return for any arbitrary member in the jth risk class is

$$s_j I_j - \sum_{i=1}^{k} s_i \alpha_{ji} L_i,$$

where L_i is the number of claims from the ith risk class and I_j indicates the claim status from the peer under consideration in the jth risk class. The actuarial fairness condition corresponds to

$$\sum_{i=1}^{k} \alpha_{ji} s_i n_i q_i = s_j q_j,$$

where $q_j = \mathbb{E}[I_j]$ is the probability of making a claim. Allocation ratios must satisfy

$$\sum_{i=1}^{k} n_i \alpha_{ij} = 1;$$

$$0 \leq \alpha_{ij} \leq 1.$$

It is clear that the Xianghubao model is based on an aggregate (non-olet) risk sharing scheme and that allocation ratios are age- and severity-specific to paying members only. In other words, allocation ratios α_{ji} are the same for all i's (the second index). For simplicity, we slightly abuse the notation to use α_j for the allocation ratio for the jth risk class. In this case,

$$\alpha_j = \frac{s_j q_j}{\sum_{i=1}^{k} s_i n_i q_i}.$$

Table 8.4 Fair mutual aid plan

Age group	Mild critical illness		Severe critical illness	
	Morbidity rate (%)	Benefit	Morbidity rate (%)	Benefit
30 days to 39-year-old	0.597	50, 000 yuan	0.733	300, 000 yuan
40- to 59-year-old	6.582	4, 531 yuan	7.273	30, 229 yuan

Note that the total cost for each month is split equally among all peers. In other words, the allocation ratio α_j is the same for all risk classes $j = 1, \cdots, k$. Therefore, the fairness condition dictates that benefit amounts should be determined by the following identity for all $i, j = 1, \cdots, k$,

$$\frac{s_i}{s_j} = \frac{q_j}{q_i}. \tag{8.2}$$

We are now ready to calculate the fair benefit amount for each group. Since Xianghubao's morbidity rates are not publicly available, we use the industry-wide data as proxies, which are based on the most recent morbidity study published by China's insurance regulator China Insurance Regulatory Commission (2013). There are four types of morbidity rates in the industry-wide data, including 6-type rates and 25-type rates for both males and females. For simplicity, we assume that half of the covered population is male and the other half female and that the population is equally spread in all ages. Then we use the average morbidity rate for 6 types of critical illness as a proxy for that of mild critical illness and the average rate for 25 types of illness as a proxy for severe critical illness.

We keep the same benefit levels for the younger age group provided in Table 8.3 and use the fairness condition in (8.2) to determine fair benefit amounts for the older age group, which are presented in Table 8.4. By industry-wide data, the morbidity rate of mild critical illness for the older age group is about 11 times that of the younger age group and the morbidity rate of severe critical illness for the older age group is about 10 times that of the younger age group. Even though Xianghubao's demographics and coverage of illness may be different from the assumptions used in this analysis, Table 8.4 offers some ballpark estimates of actual fair pricing. By comparison, benefit levels offered by Xianghubao in Table 8.3 are clearly more generous than fair to the older age group in consideration of the younger age group. This is clearly a cause of concern in long run. As the platform publishes its claims history online, younger participants may realize that they are asked to pay more than their fair share of cost and feel being taken advantage of. This finding is in fact consistent with news reports on rising policy lapses. A lack of precision in fairness is potentially detrimental not only to the profitability of one's own business model, but may also erode public trust on this nascent industry.

Unfortunately, Xianghubao was shut down in January 2022. Despite the short-lived practice, this ground-breaking business model has inspired many other innovations on mutual aid platforms in China, Japan, and other countries.

Appendix 8.A Proof of Example 8.3

Consider exchangeable random vector $\boldsymbol{X} = (X_1, X_2, \ldots, X_n)$ and the linear risk sharing rule $\boldsymbol{Y} = \boldsymbol{D}\boldsymbol{X}$, where $\boldsymbol{D} = [D_{ij}]$ is a doubly stochastic matrix. In view of the fact that $X_j \leq_{cx} X_i$ for $i, j = 1, \cdots, n$ and that $(D_{i1}X_i, D_{i2}X_i, \ldots, D_{in}X_i)$ is comonotonic, it follows from Denuit et al. (2006, Proposition 3.4.29) that for all $i, j = 1, \cdots, n$,

$$Y_i = \sum_{j=1}^{n} D_{ij}X_j \leq_{cx} \sum_{j=1}^{n} D_{ij}X_i = X_i.$$

Appendix 8.B Proof of Theorem 8.2

We make pairwise comparisons of the decentralized insurance schemes.

(i) $g^c(Y_i) \leq_{cx} Y_i$

Recall that the CAT risk pooling is given by $g^c(Y_i) = d_i \wedge Y_i + k_i + (Y_i - m_i)_+$, where $m_i > d_i$. Because $k_i = \mathbb{E}[(Y_i - d_i)\mathbb{I}_{d_i<Y_i\leq m_i} + (m_i - d_i)\mathbb{I}_{Y_i\geq m_i}] \leq m_i - d_i$, it is clear that $m_i \geq d_i + k_i$. On one hand, consider any fixed real number $y \geq d_i + k_i$. If $g^c(Y_i) > y$, then it implies that $Y_i > m_i$ and that $g^c(Y_i) = d_i + k_i + Y_i - m_i$. Observe that $g^c(Y_i) - Y_i = d_i + k_i - m_i \leq 0$. Therefore, $g^c(Y_i) > y$ implies that $Y_i > y$. Thus, $\mathbb{P}[g^c(Y_i) > y] \leq \mathbb{P}[Y_i > y]$, which is equivalent to $\mathbb{P}[g^c(Y_i) \leq y] \geq \mathbb{P}[Y_i \leq y]$. On the other hand, consider any fixed real number $y < d_i + k_i$. If $g^c(Y_i) \leq y$, we must have $Y_i \leq d_i$ and that $g^c(Y_i) = Y_i + k_i$. Note that $g^c(Y_i) - Y_i = k_i \geq 0$. Thus, we have $\mathbb{P}[g^c(Y_i) \leq y] \leq \mathbb{P}[Y_i \leq y]$. In summary, we observe that

$$\begin{cases} \mathbb{P}(g^c(Y_i) \leq y) \leq \mathbb{P}(Y_i \leq y), & y < d_i + k_i; \\ \mathbb{P}(g^c(Y_i) \leq y) \geq \mathbb{P}(Y_i \leq y), & y \geq d_i + k_i; \end{cases}$$

By the Karlin-Novikoff cut-off criterion (Theorem 2.2), we obtain that $g^c(Y_i) \leq_{cx} Y_i$.

(ii) $g^p(Y_i) \leq_{cx} g^c(Y_i)$

Recall that the cost of participant i of P2P insurance is given by $g^p(Y_i) = d_i \wedge Y_i + \pi_i$ and is bounded from above by $d_i + \pi_i$, where $\pi_i = \mathbb{E}[(Y_i - d_i)_+]$. For any fixed real number $y \geq d_i + \pi_i$, we must have $1 = \mathbb{P}[g^p(Y_i) \leq y] \geq \mathbb{P}[g^c(Y_i) \leq y]$. In contrast, for any real number $y < d_i + \pi_i$. If $g^p(Y_i) < y$, we must have $Y_i < d_i$, and that $g^p(Y_i) = Y_i + \pi_i$. It is implied that $g^c(Y_i) = Y_i + k(Y_i)$. Recall that $k(Y_i) < \pi_i$, which implies that $y > Y_i + \pi_i > Y_i + k(Y_i) = g^c(Y_i)$. Therefore, we find that

$\mathbb{P}[g^p(Y_i) < y] \le \mathbb{P}[g^c(Y_i) < y]$. Piecing two part together, we obtain that

$$\begin{cases} \mathbb{P}(g^p(Y_i) \le y) \le \mathbb{P}[g^c(Y_i) \le y], & y < d_i + \pi_i; \\ \mathbb{P}(g^p(Y_i) \le y) \ge \mathbb{P}[g^c(Y_i) \le y], & y \ge d_i + \pi_i. \end{cases}$$

Recall that by assumption $\mathbb{E}[g^p(Y_i)] = \mathbb{E}[g^c(Y_i)]$. It follows from the Karlin-Novikoff cut-off criterion that $g^p(Y_i) \le_{cx} g^c(Y_i)$.

(iii) $g^w(Y_i) \le_{cx} g^p(Y_i)$

Recall that $g^w(Y_i) = Y_i \wedge (1 - \rho_i^w)d_i + \rho_i^w d_i$. We claim that $g^w(Y_i) \le_{cx} g^p(Y_i)$. It is clear that for any fixed real number $y \ge d$, it is clear that $1 = \mathbb{P}[g^w(Y_i) \le y] \ge \mathbb{P}[g^p(Y_i) \le y]$. And for any fixed real number $y < d$, if $y > g^w(Y_i)$, it implies that $Y_i < (1 - \rho_i^w)d_i$. Therefore, we have

$$g^p(Y_i) - g^w(Y_i) = \pi_i - d_i\rho_i^w = \mathbb{E}[(d_i - Y_i)_-] - \mathbb{E}[((1 - \rho_i^w)d_i - Y_i)_-] \le 0.$$

Hence, $y > g^w(Y_i) > g^p(Y_i)$. Consequently, we must have $\mathbb{P}[g^w(Y_i) \le y] < \mathbb{P}[g^p(Y_i) \le y]$. In conclusion, we obtain $g^w(Y_i) \le_{cx} g^p(Y_i)$.

(iv) $\mathbb{E}[X_i] \le_{cx} g^w(Y_i)$

This result follows from Jensen's inequality because by assumption $\mathbb{E}[X_i] = \mathbb{E}[g^w(Y_i)]$.

Problems

1. (P2P insurance with quota-share) Consider the construction of a P2P insurance scheme for a community of n participants. Suppose that the pre-exchange risks $(X_1, \cdots, X_n)$ from participants have finite means $\mu_i = \mathbb{E}[X_i]$ for $i = 1, \cdots, n$. The aggregate risk $S = \sum_{i=1}^n X_i$ has the mean $\mu = \sum_{i=1}^n \mu_i$. The risk sharing for the ith agent is given by $h_i(s) = (\mu_i/\mu)s$. Assume that the entire community has the aggregate risk-bearing capacity of a. Any risk beyond its capacity is ceded to a reinsurer, i.e. $(S - a)_+$. Therefore, the community collectively pays the premium $\pi_i = \mathbb{E}[(h_i(S) - a)_+]$ based on the expected value premium principle.
   a. Suppose that each participant has risk-bearing capacity in proportion to their mean loss, i.e. $a_i = h_i(a)$. Determine the premium each participant pays for the ceded risk after risk exchange if each buys his/her own excess-of-loss policy.
   b. Show that the sum of participants' excess-of-loss premiums is equal to the reinsurance premium paid collectively by the community. In other words, the reinsurance premium can be allocated to participants in accordance with their excess-of-loss policies.

c. Each participant makes an initial deposit up to their risk-bearing capacity and is also eligible for a cashback if there is a remaining balance after all claims are paid. Determine the expression for their cashbacks.
d. Show that the proposed P2P insurance scheme is actuarially fair.

2. (Takaful with conditional mean risk sharing) Consider a community of participants who wish to organize a decentralized insurance scheme using a conditional mean risk sharing for its risk sharing and the mudarabah takaful model for risk transfer with a third-party takaful operator. Suppose a comonotonic conditional mean risk sharing plan is used. Denote by X_i the claim from the ith member and by $h_i(S)$ the allocation to the ith member of the takaful where S is the aggregate claim from the community. Suppose that the total contribution to the takaful fund is given by a with the ith member offering the amount $a_i = h_i(a)$. Denote the mudarabah profit-sharing rate with the takaful operator by ρ_i for the ith member.

   a. Show that the net return of the ith member is given by

   $$X_i + (1 - \rho)[a_i - h_i(S)]_+ - a_i.$$

   b. If the decentralized insurance scheme is actuarially fair, then show that each member's profit-sharing rate must satisfy

   $$\rho_i = \frac{\mathbb{E}[(h_i(S) - a_i)_+]}{\mathbb{E}[(a_i - h_i(S))_+]}.$$

3. If the underlying risk sharing rule is pro rata and the group in question is homogeneous, then the wakalah fee is the same for all participants, i.e. $\rho_1^w = \cdots = \rho_n^w = \rho^w$. Show that ρ^w can be determined by

$$1 - \mathbb{E}\left[\frac{S}{d}\right] = \mathbb{E}\left[\left(1 - \rho^w - \frac{S}{d}\right)_+\right].$$

4. If the underlying risk sharing rule is pro rata, both profit-sharing ratio and the wakalah fee are the same for all participants. Show that

$$1 - \mathbb{E}\left[\frac{S}{d}\right] = (1 - \rho^{hm})\mathbb{E}\left[\left(1 - \rho^{hw}d - \frac{S}{d}\right)_+\right].$$

Chapter 9
DeFi Insurance

Decentralized finance (DeFi) is an emerging market practice for providing financial services through blockchain technology without traditional market intermediaries such as financial institutions. DeFi insurance is a class of DeFi protocols that provide insurance coverage for crypto assets against smart contract risks. Like many DeFi protocols, there have been new practices developed for the decentralization of various components of the traditional insurance ecosystem. This chapter starts with a brief review of key concepts for blockchain technology, provides an overview of DeFi protocols, and ends with a discussion of current practices in DeFi insurance.

9.1 Elements of Blockchain Technology

Most of decentralized insurance models discussed in this book have been implemented in practice with various types of infrastructure. Historic risk sharing organizations, such as burial societies in ancient Rome, guilds in Middle Age Europe, and fraternal benefit associations in many parts of the world, have long existed before the development of modern Internet technology. The accounting of financial transactions is done through book-keeping with paper and pencil. These mutual support organizations are typically voluntary and self-organized around a shared ethnic background, religion, socio-economic status, occupation, geographical region, or other bases. While these organizations provide broader social support beyond financial means, their core mechanism for financial support can be characterized as decentralized insurance. In modern days, many decentralized insurance models, such as online mutual aid, peer-to-peer insurance, and takaful have been marketed and distributed using mobile apps or Internet platforms. These new telecommunication technologies extend the reach of decentralized insurance to much larger networks of geographically and demographically diverse population, which would not be possible in old

R. Feng, *Decentralized Insurance*, Springer Actuarial,
https://doi.org/10.1007/978-3-031-29559-1_9

times. It is fair to say that decentralized insurance has evolved with the advancement of technology.

In recent years, there has been an emergence of new decentralized business models with the rise of blockchain technology. While the discussion of actuarial principles in this book is largely independent of hardware infrastructure, we shall nevertheless briefly introduce fundamental concepts of blockchain technology, which can help us understand the unique opportunities brought about by this new technology for decentralized insurance. Readers who are interested in technical details are referred to Lipton and Treccani (2021).

9.1.1 Promise of Smart Contracts

Distributed ledger technology, also known as blockchain, allows economic transactions to be recorded and stored in a verifiable and immutable way and synchronized in a network of participants. A blockchain uses a data structure that packages transaction data in cryptologically protected blocks and adds to an ever-growing chain of blocks.

A smart contract is an application that sets a protocol on blockchain to execute contracts between business parties with little to no human intervention. In other words, business logic in traditional contracts is coded into self-executing scripts that perform on blockchain platforms. An ecosystem built with smart contracts can facilitate business transactions without centralized authorities and provide automated and trustworthy marketplaces.

Smart contracts are in many ways technology solutions to support the ideals of decentralized insurance.

1. Smart contracts permit trusted transactions to be carried out among anonymous participants without a central authority, a legal enforcement, and human intermediary. Unlike centralized systems, where transactions are facilitated by a central authority, smart contracts rely on a network of anonymous participants to corporate in the maintenance of records.
2. Smart contracts enable direct peer-to-peer contingent payments. As an intermediary in traditional insurance, an insurer uses premiums from some policyholders to pay for others' benefits upon contingent risks. Such redistribution of money takes place behind the scene, because insurers operate on a large bulk of policies. Policyholders can only observe their bilateral relationship with an insurer and hence are largely unaware of the opaque underlying risk sharing mechanism. In contrast, smart contracts written by a network of participants permit digital assets to be transferred from one user to another upon some agreed contingencies without the need for an intermediary. Such transactions are coded in public ledgers and fully auditable.
3. Smart contracts achieve cost efficiency through the automation of contract conditions. Blockchain-based technologies have been applied in the insurance industry

with examples such as claims processing, on-demand insurance, and automated underwriting (c.f. Gatteschi et al., 2018; Hans et al., 2017). Smart contracts are perfectly suitable for parametric insurance solutions. For example, a flight delay insurance product was launched by AXA in 2017 based on blockchain technology. The smart contract is connected to global air flight databases and triggers payments to policyholders when an over two hour delay is detected in flight tracking databases. There is no need for human insurance adjusters or appraisers.

In the following subsections, we shall provide a brief overview of the architecture of blockchains. Readers may also skip the background content, as the rest of the chapter does not directly depend on an understanding of blockchain architecture.

9.1.2 *Asymmetric Key Cryptography*

At the core of blockchain technology is asymmetric key cryptography. The technical details of cryptography require some knowledge of number theory, which is beyond the scope of this book. We shall only focus on the logic behind the infrastructure rather than getting into the technicality.

The purpose for the use of cryptography in the context of blockchain applications is to safeguard the interests and ownership of intended users in financial transactions against malicious acts of adversaries. Encryption is a process of converting an original human-readable digitized message, called *plaintext*, into unintelligible codes, known as *ciphertext*. The reverse process of turning ciphertext back to plaintext is called decryption. Cryptographic keys are essential inputs for both encryption and decryption. The sender of the message only shares the cryptographic key with an intended user. The encryption technology is intended to make it hard to break the encryption without the knowledge of the key.

In modern cryptography, encryption and decryption are based on sophisticated one-way mathematical functions, for which function values are easy to compute given known inputs (cryptographic keys) and for which inverse functions are very difficult to find. The early encryption technology, known as *symmetric key cryptography*, requires both a sender and a receiver to use the same key. Figure 9.1 shows the process of encryption and decryption with symmetric key cryptography. Most historic encryption techniques use symmetric key cryptography. Many techniques based on

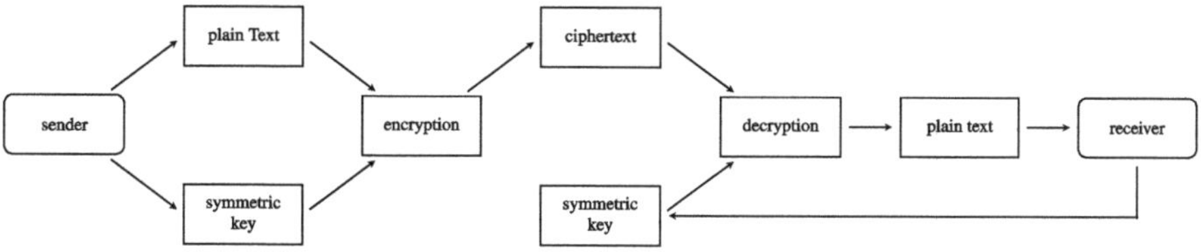

Fig. 9.1 Symmetric key cryptography

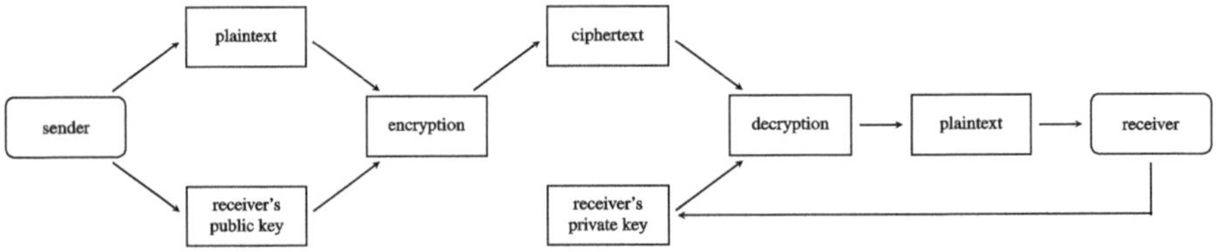

(a) Encryption-decryption paradigm

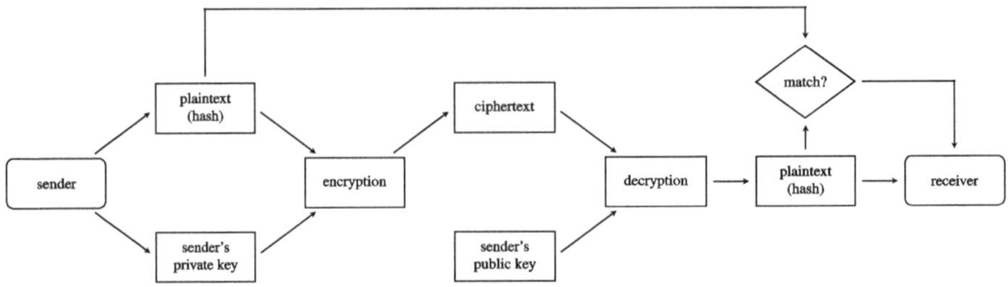

(b) Digital signature verification

Fig. 9.2 Uses of asymmetric cryptography

symmetric key cryptography are still used today for a wide range of commercial applications. However, a disadvantage of this technology is that each pair of parties for communications require a distinct key and both parties have to keep the key secret and secure, which create difficulty for key management in the presence of a large number of users.

Many of today's industry applications are based on *asymmetric key cryptography*, which is a new technology to use a pair of mathematically related cryptographic keys for communications, a *public key* and a *private key*. The originator of any transaction should keep her own private key secret and safe. In contrast, her public key is disclosed and shared with others through publicly accessible directory. It is mathematically very difficult to infer a private key from its corresponding public key(s). However, asymmetric encryption is considerably slower to compute than symmetric encryption. That's why symmetric cryptography is still commonly used for low and moderate security communications whereas asymmetric cryptography is used for high security transactions.

There are two common uses of asymmetric key cryptography that are building blocks of distributed ledger technology.

1. Encryption-decryption
   A receiver's public key is used for encryption and the receiver's private key is used for decryption. The encryption-decryption paradigm is shown in Fig. 9.2a. The sender of a message takes a plaintext and the intended receiver's public key as inputs for encryption, which generates a ciphertext. The receiver uses the ciphertext and his own private key as inputs for decryption and subsequently

obtains the original plaintext as an output. Even if an adversary would to intercept the ciphertext, he would not be able to break the encryption and understand the message without the receiver's private key.

2. Digital signature verification
   A sender's private key is used to sign a message and the sender's public key is used to verify the signature. The digital signature verification process is shown in Fig. 9.2b. Signing a message means that a sender uses own private key and the message (plaintext) as inputs to an encryption algorithm and generates a *digital signature* (ciphertext). The receiver can use the sender's public key to decrypt the ciphertext. If the receiver obtains a plaintext, it indicates that only the sender can be the one who signs the message. Note that it is impractical to encrypt large amounts of data with an asymmetric key which is computationally demanding. Therefore, in practice, a hash function, which maps text of any length to a hash value (ciphertext) of a fixed length, is used and the sender signs the hash value instead of the original text. The hash value and the signed text are then passed on to the receiver as inputs for decryption with the sender's public key. The decrypted text is then compared with the hash value to see if they match. If they do, it indicates that the text was created by the sender.

9.1.3 Hash and Merkel Tree

Hash function is a type of mathematical function that is easy to compute but hard to find a preimage for. A hash function maps a message of any size to a ciphertext of 256 bits. Such a function is not invertible, as it is a many-to-one function. A useful hash function is collision-resistant, which means that it is hard to find two different inputs that produce the same output through the hash function. For any given hash value, one can only find a preimage by brute-force random guess, which is expected to be very time-consuming. The hashing is different from encryption in the sense that the process is irreversible and the ciphertext is not intended to be decrypted to the original plaintext.

One of the most well-known Hash function is SHA-256, designed and published by the United States National Security Agency. The abbreviation SHA stands for Secure Hash Algorithm and the number 256 indicates the size of its output irrespective of the size of inputs. For example, we can hash the plaintext

this is a book

to generate the following hash value in hexadecimal.

255c905b3539fe67578d497a13883ab20d8004024ea476545608ef131f03f4a9

We change the plaintext by adding a stop sign to

this is a book.

Then the hash value is changed beyond recognition from that of the plaintext without the stop sign.

df32105c74aaa79b0ebd623ed3ee288dfc9c4065c000d655f2ca0e4db7a3b994

Hashing is used at least for two different purposes for blockchain technology, one of which is to compress a large volume of data into a small amount of data. Because there is a huge amount of transaction information to be registered in each block, it is impractical to store the complete information in a block. A *Merkel tree*, or a hash tree, is a data structure used in blockchain for compression and verification. Figure 9.3 visualizes the structure of a Merkel tree. The boxes at the bottom represent the data packages A_0, A_1, A_2, A_3 containing transactions. They are hashed to generate the first layer of hash values, M_0, M_1, M_2, M_3. Each pair of hash values from the first layer are appended and hashed to generate a new hash value in the second layer. The pattern continues until a single root hash value is generated, which will be inserted into the header of a block when a block is built. It is clear that the procedure has a tree-like structure with each leaf representing a hash of data underneath it.

The Merkel tree offers a digital footprint of a set of transactions and a highly efficient way to verify a particular transaction. For example, if a transaction were to be recorded in the data package A_2, then, to prove it is authentic and has not been tampered with, one would just need to send A_2 and hash values M_3, M_{01} to be verified instead of the entire data sets A_0, A_1, A_2, A_3. In Fig. 9.3, we show the required hash values in thickened boxes in order to verify the authenticity of the data package A_2. Note that verifying the hash values with a data package is computationally much more efficient than combing through all historic data ever recorded on a blockchain.

The second use of hashing is mining in a proof-of-work consensus for block building, which we shall discuss in the next section. In a nutshell, the process of

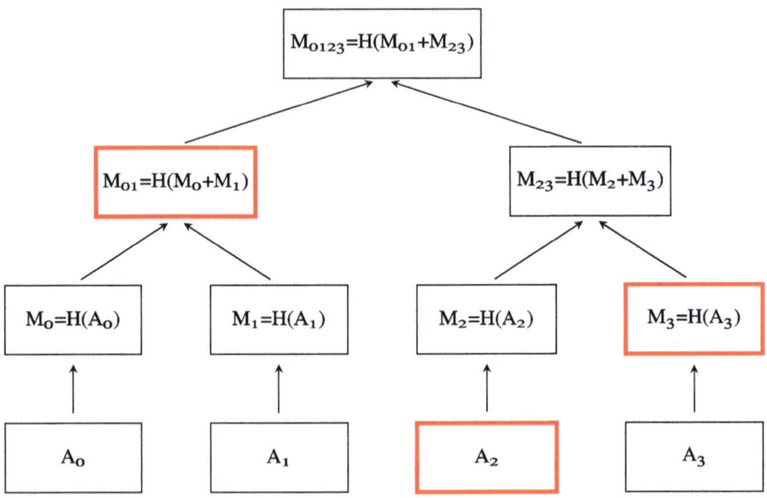

Fig. 9.3 Merkel tree

Table 9.1 Adjusting nonce to meet a difficult level

Input	Hash value
Book	92719fe0cf8cd51592af31ee8a5736d79f7273777fa3f7b70bfe993a4cd32180
Book1	c676562512aa31205cee48429b9e8de069823d3f1e33d87db981dd20d32ef464
Book2	3cbcece7185ad3f34b6580c5342231d5c407c10b787287d95d8652170d55cbb5
Book3	cb779dce490a299d6196b3870ca0978bcf16851b2fa4a12f037e5f11c13ea4fa
Book4	43c108d70fb9fe7aa00289c615b4d8b3e92f60c366df91cd4707395aebe7e717
Book5	bb6e774039554e1b5a0187ac28c6bfd66c1095d02202c0adae2d0ddcbf6a7c9b
Book6	c05173338a798ab3f102f41812c6724532077aed0a0af832de3799348f698202
Book7	917c3c5c08c50791d5fc35c67d349e351a3dfbc4683b290f61e1b0dcc5cf73df
Book8	053741156d0d6631b9a8afccfa75c7326844db3710b049d25b7d6d63ea0b2c6c

mining is to look for 4-bit inputs, known as *nonces*, that can generate a hash value below a certain target level. Nonces are numbers only used once for each block. In order to build a block, a computer node first packages a set of transaction data with some nonce using a Merkel tree structure and then appends a block header with another nonce. This structure of a block with two nonces is shown in Fig. 9.4. Observe that there is a nonce at the bottom layer of a Merkel root and another nonce in the block header. The discussion of block structure is deferred to the next subsection.

Bear in mind that it is usually very time-consuming to random guess inputs of a hash function in order to obtain a desired hash value. It is even harder to find the two nonces in a blockchain structure that generates a hash value below a target level. To illustrate how difficult it is to find a nonce, we consider appending the text *book* with a nonce to generate a hexadecimal number that begins with a 0. Table 9.1 shows that it takes 8 iterations to generate the initial number 0. The actual hashing puzzle for block building requires many more zeros and nontrivial initial numbers with two nonces. One can imagine how time-consuming it is to meet such a difficulty level.

Even though the desired nonces are not unique and in theory a computer node can find them through brute-force exhaustion of numbers, it requires an enormous amount of computational work and by chance to be the first one to find such a pair of nonces. This is precisely the reason for a proof-of-work consensus mechanism, which requires a computer node to show an evidence of work before it can be recognized as an originator of a data block.

9.1.4 Structure of a Blockchain

Each blockchain is structured differently. Here we use Bitcoin as an example, as its structure is relatively simple. A blockchain relies on a network of computing nodes to maintain a distributed ledger of all transactions. As in any decentralized network with no central authority, a decision-making, the building of new blocks in the case of blockchain, can only be achieved through some consensus mechanism to avoid any miscommunication or malicious attacks to the system.

The proof-of-work is a consensus mechanism in which a participant proves to others that a certain amount of work has been performed and in which the participant with the proof is charged to add a new blockchain to a chain. As discussed in the previous section, the proof is given by the nonces to generate a hash below a certain target. The process of looking for the right nonces is called *mining*. Because it requires heavy computational efforts to solve a hashing puzzle, it is unlikely for more than one computer node to find a proof at the same time. Once the puzzle is solved, the computer node broadcasts the proof to the entire computing network and it is very easy for other nodes to verify the hash result. The computer nodes that aim to mine new blocks are also referred to as *miners*.

As the computing network is decentralized and spread out around the world, it may occur that multiple computer nodes all find their proofs over a short period of time. Due to the unpredictability of mining and the broadcast speed of the network, other computer nodes may disagree on which block from a miner with proof to be added to the currently agreed upon blockchain. This situation is called a *fork*, where different blocks are added to the end of current blockchain. The proof-of-work mechanism resolves this issue by having computer nodes to wait for up to 6 blocks. Computer nodes adopt the longest chain (the highest accumulated difficulty level) as the current chain. All transactions on the longest chain are considered by computing nodes as part of the valid history and other transactions are not considered accepted. For every successful block, the miner receives a mining reward in bitcoins and transaction fees from users who request their transactions to be recorded in new blocks. The average mining time for each block is 10 minutes and hence it takes around one hour to generate 6 blocks. All mining activities are driven by financial incentives. It makes no sense financially for a miner to keep working on a shorter chain. Hence, all miners shall eventually agree on the current status of the blockchain. An exception is a rule change, in which the community of miners agrees to split a chain into different permanent versions of a blockchain. Such an example is the split of Ethereum and Ethereum Classic chains in 2016 after the community disagreed on whether to rollback blockchain records after a hack to a critical protocol on Ethereum at the time.

Each Bitcoin block can be visualized as in Fig. 9.4. The block header contains information on the version of the Bitcoin protocol, the hash of a previous block header, the Merkel root of all transactions in this block, a timestamp, the target difficulty, and nonces. The blocks are connected through hash pointers which are hash values of previous blocks. The main body of a block is the transaction data that a miner pulls from a pool of undocumented transactions waiting for inclusion. The first transaction within each block is called the coinbase transaction, which documents how much rewards and fees should be credited to a minder. The block also contains the addresses of senders and receivers, the amount of bitcoins in each transaction, the private key signatures authorizing the transactions, and timestamps.

The distributed ledger, which is a collection of data blocks, represents the system memory of all transactions since the beginning of the blockchain. After all computer

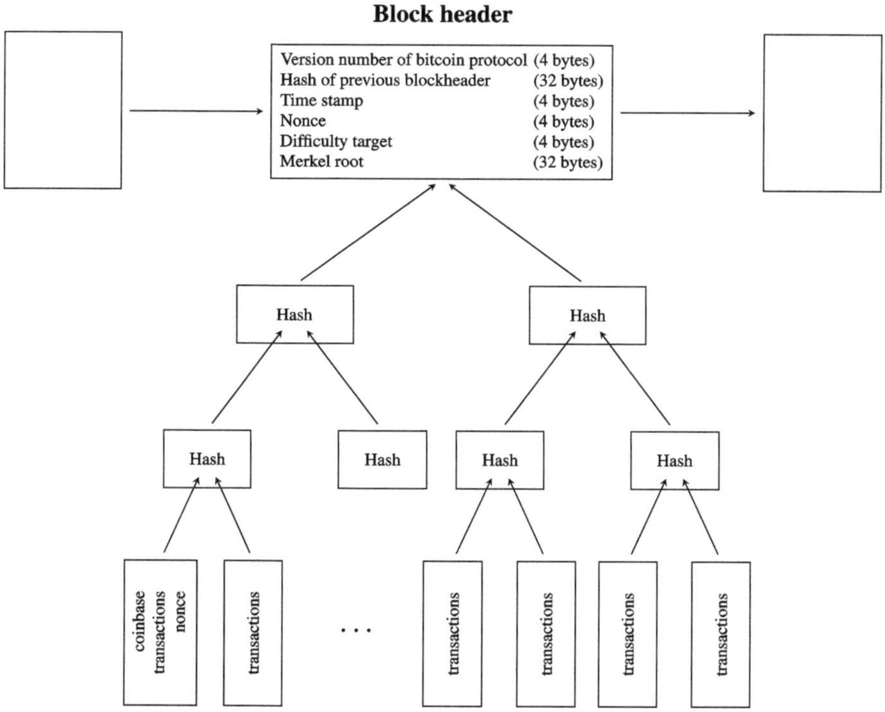

Fig. 9.4 Structure of Bitcoin blockchain

nodes reach an agreement on the history of blocks, each computer node on the blockchain network has exactly the same copy of the ledger. All transactions are publicly available and open for audits. The blockchain is often said to be immutable, because any change in a transaction would alter all subsequent data. Because all other computing nodes are continually expanding the current known longest chain, one has to acquire 51% of the computing power on the network to be able to build another fork with a new longest chain. This is referred to as the 51% *attack* problem. While this problem exists in theory, it is practically impossible for someone to acquire such a level of computing power on a large network of computer nodes.

The proof-of-work mechanism is often criticized for its intensive computation of mining, which requires computer nodes around the world to consume a great amount of electricity and is hence considered environmentally unfriendly. *Proof-of-stake* is used by the industry as an alternative consensus mechanism. In contrast with computational competitions in the proof-of-work, the proof-of-stake mechanism reduces computational efforts by randomly selecting computer nodes, called *validators*, to verify transactions and build new blocks. Each validator has to lock up a minimal amount of cryptocurrencies, called *stake*, in a smart contract. If a validator is caught validating fraudulent transactions, he may lose some or all of his stake as a penalty. The selection of a validator varies greatly by blockchain and often depends on the amount of stake and the validator's due-diligence history. Proof-of-stake has its own drawbacks. For example, as it is not computationally heavy, the cost is also low to bribe others to reach a consensus on fraudulent records. Validators with large amounts of stake may have an oversized influence on a proof-of-stake blockchain. This is the reason why many blockchains, such as Ethereum, start with a proof-of-work consensus and transition to a proof-of-stake consensus when the network of computing nodes reaches a sufficiently large size.

9.2 DeFi Ecosystem

Decentralized Finance (DeFi) refers to the technology, the process, and the infrastructure developed to provide financial services without central authorities. In everyday language, the two terms *decentralization* and *disintermediation* are in fact different. The former refers to the process in which the control of a central authority is transferred to multiple entities, whereas the latter means the reduction in the use of intermediaries between consumers and providers. In the context of financial industry, the two terms are often intertwined because, when intermediaries, typically established financial institutions, are removed, the technology by itself does not require any permission from central authorities, such as market regulators. Therefore, decentralized finance has been used as all encompassing term for all blockchain-based financial applications.

Decentralized applications (Dapps) are computer programs that run on a distributed computing system. When the distributed computing system is used to record and store financial transactions, it is often referred to as distributed ledger. Dapps have been used to perform financial functions and hence are also referred to as *smart contracts* or *DeFi protocols*.

Smart contracts have value because they have been programmed to trade and store digital currencies and assets. *Collateral* has to be locked into smart contracts in order for investors to participate in financial transactions in the DeFi space. Therefore, the cumulative collateral locked in smart contracts is often used as a measure of market size and referred to as *Total Value Locked* (TVL).

To help understand the significance of various applications to be discussed in this chapter, we shall present some market data. However, these statistics should only be interpreted in relative terms, as the market evolves at a fast pace. Some statistics may already be outdated by the time a reader reaches this book. Readers who are interested in the latest market data are referred to https://defillama.com/ and https://www.coingecko.com.

The earliest successful financial application of a blockchain is a peer-to-peer payment system based on a distributed ledger, Bitcoin, which was created in 2009 by a presumed pseudonymous person named Satoshi Nakamoto. Since then there have been thousands of public blockchains that became available and compete for customers. The landscape of DeFi protocols has evolved at a very fast pace. Top-ranked blockchains with the highest TVL in smart contracts include Ethereum, Binance

Fig. 9.5 Total value locked in DeFi protols. *Source* https://defillama.com/

Smart Chain (BSC), Tron, Avalanche, Solana, and Polygon. The Ethereum platform, launched in 2015, is the first blockchain developed with smart contract functionality and by far the chain with the highest market capitalization accounting for more than half of the TVL in the DeFi space. Figure 9.5 shows the evolution of the DeFi market in terms of TVLs, which started in late 2018 with the emergence of the first major decentralized exchange (Dex), Uniswap, on the Ethereum chain. The DeFi market saw an exponential growth in 2021 with the TVL peaking at more than $180 billion in November 2021. The DeFi space experienced a steep and broad market decline in all major cryptocurrencies with the collapse of algorithmic stablecoins, Terra Luna, in late April 2022, which shall be discussed in the next section. The Terra Luna crash led many crypto lending firms such as Celsius and Voyager Digital to file for bankruptcy, causing the climate for crypto investment to deteriorate. In November 2022, the world's third largest crypto exchange by volume, FTX, which was once valued at $32 billion, collapsed over a 10-day period. It started with a news report raising concerns over its undisclosed leverage and solvency issues. Its rival exchange Binance sold its entire position on FTX's native tokens, leading to FTX's liquidity crisis. After a failure to secure a bailout from a series of investors, the exchange filed for bankruptcy. The FTX fiasco sent shock waves throughout the crypto market. It is estimated that over one million investors lost money subsequently. As of January 2023, the TVL in the DeFi space has shrunk to only about a quarter of its peak value in November 2021.

The DeFi industry has organized itself into a massive ecosystem consisting of many interacting organisms. There are different classifications of sectors within the DeFi space. Figure 9.6 shows a breakdown by DefiLlama including decentralized exchanges, lending, bridge, collateralized debt position, liquidity staking, yield farming, financial derivatives, stablecoins, insurance, and other services. As of August 2022, the collective TVL of DeFi insurance protocols is only around 500 millions,

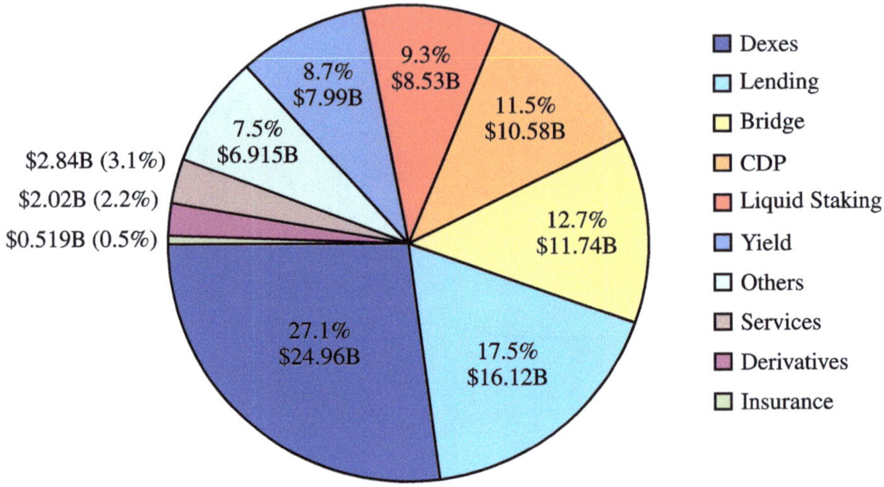

Fig. 9.6 Breakdown of TVLs in sectors. *Source* https://defillama.com

which is about 1% in the DeFi market. The global total asset under management in the financial industry is about $94 trillion in 2017, whereas the total assets of insurance companies worldwide are around 35 trillion dollars. If the traditional industry offers a glimpse of what the DeFi industry may look like over the long term, then there is a huge room for growth for DeFi insurance.

The three most regulated and fundamental sectors in the traditional financial industry are securities, banking, and insurance. We shall discuss their counterparts in this chapter, namely decentralized exchanges, lending, and insurance. While the focus of this chapter is on decentralized insurance, we shall also briefly describe a few key protocols in the other two sectors with which decentralized insurance applications often have interactions. Bear in mind that, while these protocols all provide different utilities in the DeFi space, they often share very similar logic for decentralization.

9.2.1 Stablecoins

While government-issued currency, also known as fiat currency, is considered as a unit of denomination and a medium for the storage and exchange of value, the value of a cryptocurrency is often too volatile to be used as a medium of exchange. For example, in May 2010, a Florida resident bought two Papa John's pizzas for 10, 000 bitcoins, which was the world's first transaction to buy physical goods with Bitcoin. In today's price, the person spends about $200 millions to purchase the two pizzas, which are arguably the most expensive pizzas in the world. Statistics show that Bitcoin prices can go up by as much as 16% and down by as much as 18% on the same day. If one were to use Bitcoin to purchase a commodity, the person may suffer huge losses between the time at which the price is agreed upon and the time at which the trade is executed. Therefore, many argue that cryptocurrencies should be considered as digital assets rather than currencies. Stablecoins are created precisely to address such issues and to be used as currencies.

In simple terms, a *stablecoin* is a cryptocurrency with a stable value. Usually, it is pegged to the value of some fiat currency or commodity such as gold. For example, a currency called USTD or Tether is exchanged with US dollar one-to-one on the Ethereum blockchain. However, stablecoins only have a short history and there are very few places outside the crypto space that accept stablecoins for transactions. Stablecoins have a long way to go before they truly become a currency for the general public. Stablecoins are mainly used for two purposes in the DeFi market.

- Investors use stablecoins to reduce investment risk. When the value of a cryptocurrency fluctuates violently, investors can convert them to stablecoins on an exchange to preserve their asset values. When the investors decide to revert to speculative activities, they can convert these stablecoins back to cryptocurrencies.

- Stablecoins can be transferred quickly. Many exchanges cannot directly convert Bitcoin into fiat currencies due to regulatory reasons. It is a common practice to use stablecoins to replace fiat currencies. It is easier and faster to transfer cryptocurrency assets between various exchanges than fiat currencies. Transfers between exchanges take seconds, or up to an hour, compared to days for cross-border wire transfers with traditional banks.

There are two commonly used methods by which stablecoins are designed to maintain their values in the DeFi industry.

1. Collateralized stablecoins

   Like in the traditional financial industry, collaterals can be used to build market trust. When a user buys a stablecoin, the issuer of the stablecoin deposits fiat currency of equal value with a bank. The advantage of such an approach is that the total amount of the stablecoin circulating in the market should be the same as the total amount of fiat currencies deposited by the issuer. The disadvantage is that the underlying assets are frozen and cannot be invested. It is also not easy to inspect the deposit by ordinary investors. Companies that hold fiat currencies are often subject to government regulation to perform know-your-client (KYC) or anti-monetary-laundry (AML) identity verification on users who purchase stablecoins.

   There are stablecoins that use cryptocurrencies as collaterals. For example, another very common stablecoin, DAI, uses ether as a collateral asset. The deposited assets of DAI are recorded on a blockchain, and investors can verify the transactions on a block explorer, where the public can access details of any transaction on the blockchain. The disadvantage of the crypto collateral is that the value of these crypto assets fluctuates wildly. MakerDAO, the issuer of

DAI, requires 150% of Ether to be collateralized in exchange for DAI, in order to prevent the price from falling too fast and the fund balance becomes deficient. Collateralization is the most commonly used method for creating stablecoins in today's crypto market.

2. Algorithmic stablecoins

   An alternative approach is to adjust the supply of a cryptocurrency in the market to stabilize its value. A smart contract algorithm can be used to adjust the amount of currency in circulation. The price goes up when a lot of people buy the currency. Then the algorithm increases the supply, bringing the price back down to $1. Similarly, the price goes down whenever many people sell the currency. Algorithms reduce the supply of this currency and allow the price to rise to $1. This type of smart contract is not backed by any asset. The issuer effectively plays the role of a central bank.

   It is unclear if this approach would work in the long run. Because the pegging method by supply management has been tried by even the most powerful governments in the world and failed many times in history. For example, the US dollar had once been pegged to the price of gold under the Bretton Wood system since 1944, and the system was eventually abandoned in 1971.

The top five stablecoins by market capitalization are Tether (USDT), USD Coin (USDC), Binance USD (BUSD), DAI, and True USD (TUSD), all of which are collateralized. The largest algorithmic stablecoins by market capitalization are USDD, Neutrino USD (NUSD), and Fei USD (FEI). The smaller market capitalization of algorithmic stablecoins is attributable to the crash of TerraUSD (UST) in May 2022. The algorithmic supply management was carried out by exchanging UST with a native coin called Luna on the Terra blockchain. When the UST fell below its dollar peg, the price of Luna also plummeted and the mechanism of supply management through minting and burning failed to deliver incentives to validators and hence the market lost confidence on its ability to stabilize. The selling pressure on both UST and Luna eventually broke the algorithmic peg. The incident caused the entire crypto market to wipe out about two-thirds of its capitalization in 2022.

It should be pointed out that, collateralized or algorithmic, stablecoins are not truly decentralized. They merely replicate the role of traditional banks on blockchains. The main role of stablecoins is the same as that of legal tender. Yet they are not directly controlled by financial market authorities. It is unclear whether regulators around the world will allow the continued existence of stablecoins in the future. Many argue that stablecoins are a stepping stone for the industry to move to completely decentralized cryptocurrencies. Some believe that, when cryptocurrencies reach trading volumes comparable to fiat currencies, their values would stabilize and there would be no need for stablecoins.

9.2.2 *Decentralized Exchanges*

There are tens of thousands of cryptocurrencies in the market today, each with different functionalities and values. A *native token* refers to a cryptocurrency that is used as base token of a blockchain to pay transaction fees. For example, Ether, or ETH in exchange symbol, is the native token of the Ethereum blockchain. A *non-native token* is a cryptocurrency created for a particular service provided on a blockchain ecosystem. Stablecoins are examples of non-native tokens. Cryptocurrency exchange is a critical infrastructure in any blockchain ecosystem for buying and selling cryptocurrencies.

There are two types of trading mechanisms in the DeFi market. One is *centralized exchange* (CEX) that uses many of the same trading mechanisms as traditional financial institutions. The other is *decentralized exchange* (DEX) where cryptocurrencies are traded through smart contracts between buyers and sellers.

Centralized Exchange

CEXs usually use order books to facilitate the trading between buyers and sellers. An order book is used, especially in a stock exchange, to record orders placed by buyers and sellers for assets. Simply put, it is used to track whether the price specified by a buyer matches that by a seller. If one limits the price at which to buy or sell, it is called a limit order. If one is willing to accept the best price in the market, it is called a market order, in which case both parties close the deal and the order is cleared from the ledger. If the prices do not match, the order is queued on the ledger and the buyers or sellers wait until new orders to meet the specified prices.

CEXs have operated for over ten years. Coinbase and Binance are the most well-known CEXs for cryptocurrencies. Centralized exchanges are similar to traditional exchanges with employees and physical locations. However, they are regulated where they are registered and hence their customers are subject to the identity authentication of KYC and AML, which are considered an invasion of privacy by some investors. Since they are regulated, these exchanges can be held accountable for their financial transactions and have more institutional collaborators. Like typical businesses, CEXs provide customer service, and they have more communication channels for customers. For investors, especially beginner investors, they have a better user experience. There is less variety of cryptocurrencies listed and traded on these exchanges because cryptocurrencies have to be vetted by the exchanges. These cryptocurrencies are less prone to scams and frauds.

However, CEXs are often criticized for their use of *custodial trading*. Customers have to first transfer cryptocurrencies from their wallets to that of an exchange. The trading occurs in the database of the exchange and is not recorded on chain. Customers regain the control of cryptocurrencies only after withdrawals from the exchange. If required by the regulatory authorities, a CEX can in theory confiscate their customers' assets. Or, if a centralized exchange is hacked, it can halt trading and prevent customers from accessing their deposits.

Decentralized Exchange
A wide variety of cryptocurrencies are traded on DEXs. The trading volume on DEXs such as Uniswap is very close to that of CEXs like Coinbase. One can find virtually any kind of cryptocurrencies, as issuing cryptocurrency on a DEX is done through a decentralized process. DEXs often use *non-custodial trading*, which means customers retain custody of their cryptocurrencies and are responsible for managing their own wallets and private keys. All footprints on DEXs are completely public and on the ledger. One can review the past transaction records of the exchange at any time and anywhere. DEXs are not subject to any regulation and can be used by anyone anywhere in the world without the need for KYC identity checks.

DEXs do not have physical offices or formal employees. All transactions are automated through smart contracts. There is always a community of programmers who collectively maintain the functioning of the protocol. The ultimate goal of a decentralized exchange is to hand over the control of the exchange to holders of governance tokens. There is an expectation for transitions from semi-decentralization to full-decentralization over time. DEXs rarely provide data tracking or chart analysis. Users need to find some techno analysis tools by themselves to help make investment decisions. DEXs do not have customer services. Some exchanges have their own communities through Telegram or Discord, in which there are a lot of members who can help new investors with problem-solving. Because there are channels of communications between members on a decentralized exchange, there are also scammers who try to defraud investors. There is no entry from fiat currency to DEXs. Therefore, investors need to purchase cryptocurrency on other platforms in order to exchange them for other cryptocurrencies. Uniswap and SushiSwap are the two best-known DEXs. Figure 9.7 offers a snapshot of transactions through the Uniswap protocol.

A disadvantage of a DEX is that it only allows transactions between cryptocurrencies on the same blockchain. For example, only Ether, the native coin of Ethereum,

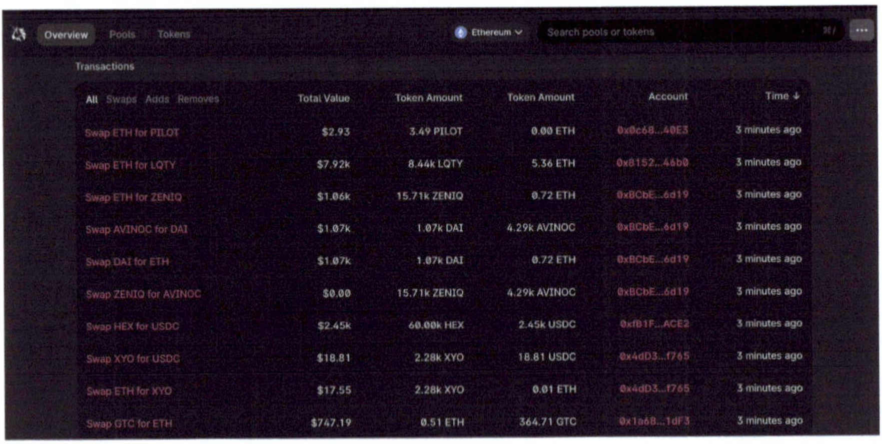

Fig. 9.7 A snapshot of Uniswap transactions

and other tokens created on the Ethereum blockchain can be traded on Uniswap. However, technical solutions have been developed to address cross-chain transactions. If we want to exchange cryptocurrencies on different blockchains, we could use a blockchain bridge.

Automated Market Maker

Most DEXs use Automated Market Maker (AMM) algorithms to facilitate the trading of cryptocurrencies. An order book system is in essence a peer-to-peer match-making process where an order from a buyer waits to be matched with an order from a seller. In a market with limited liquidity, it may take a long time to find a match. In contrast, the AMM system is a peer-to-smart-contract process where a customer swaps tokens instantly against a liquidity pool at a price determined by an algorithm. A liquidity pool is a reserve for a pair of available tokens. Liquidity pool reserves are provided by investors, often referred to as liquidity providers. Liquidity providers are incentivized to earn transaction fees, which are charged on all token swaps on a DEX. This process is also called liquidity mining.

A common AMM algorithm works by maintaining a constant product of the amount of liquidity on both sides of the pool. Take a ETH-DAI trading pair as an example.

$$\text{ETH Liquidity} \times \text{DAI Liquidity} = \text{Constant}.$$

Suppose that there are 100 ETH and 20, 000 DAI in the liquidity reserve. Then the constant is given by

$$100 \times 20,000 = 2,000,000.$$

A customer who wants to buy 1 ETH using DAI may send 202.02 DAI to the DEX smart contract address in exchange for 1 ETH. This is doable because the liquidity pool is left with 99 ETH and 20, 202.02 DAI. In other words, the price is set in such a way that the pool maintains its constant product, i.e.

$$99 \times 20,202.02 = 20,000,000.$$

This mechanism is formulated in mathematical terms. In early versions of Uniswap, this constant factor AMM algorithm is determined by

$$xy = k,$$

where x and y are the reserves of two cryptocurrencies and k is a constant. While this is easy to implement, this algorithm is capital inefficient. Because cryptocurrencies are expected by the market to trade within a reasonable range of prices, only a fraction of the reserves is actually used in the pool.

To address this issue, Uniswap introduced in Adams et al. (2021) a new AMM with the feature of concentrated liquidity. Let us denote by p the price of a swap between two cryptocurrencies with reserves x and y and represent the price as the

Fig. 9.8 Reserve

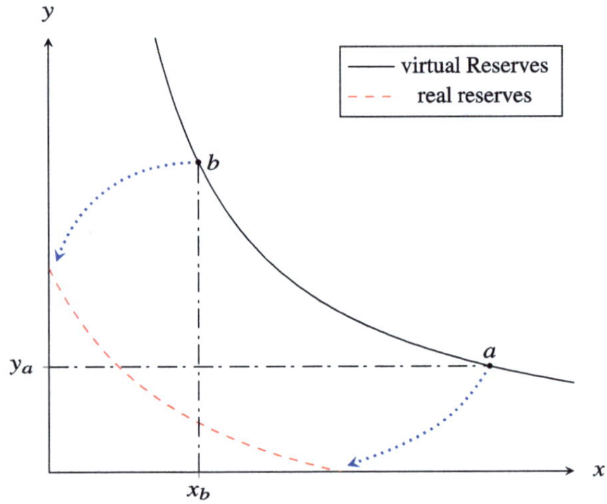

ratio $p = y/x$. The new algorithm is based on the idea that the pool only needs enough capital so that the price stays within a specified range, say $p_a \leq p \leq p_b$. To maintain the same curve as in the early versions of Uniswap, the reserves for two cryptocurrencies are now considered virtual reserves. Denote the virtual reserves for the two cryptocurrencies to reach the lower bound of the price range by x_a and y_a and those to reach the upper bound of the price range by x_b and y_b. It is clear that the virtual reserve x_b can be determined by

$$\frac{y_b}{x_b} = p_b, \qquad x_b y_b = k, \tag{9.1}$$

which implies that

$$x_b = \sqrt{\frac{k}{p_b}}.$$

Note that the virtual reserve is at the lowest level x_b for the first cryptocurrency when the price $p = y/x$ reaches the high end of the price range p_b. Similarly, the virtual reserve y_a is determined by

$$\frac{y_a}{x_a} = p_a, \qquad x_a y_a = k,$$

which gives the solution

$$y_a = \sqrt{k p_a}.$$

The virtual reserve is at the lowest level y_b for the second cryptocurrency when the price $p = y/x$ reaches the low end of the range p_a. Denote the real reserve by x and y for the two cryptocurrencies, respectively. We replace the virtual reserves in (9.1) by $x + x_a$ and $y + y_b$ and arrive at the new pricing curve of Uniswap version 3

$$\left(x + \sqrt{\frac{k}{p_b}}\right)\left(y + \sqrt{kp_a}\right) = k.$$

In this new version, liquidity providers can set their own price range around the current market price of cryptocurrency pairs and reduce the cost of capital by concentrating on a narrow range of reserves.

9.2.3 Decentralized Lending

Lending and borrowing are fundamental services of the traditional banking industry. Banks accept deposits from lenders and make loans to borrowers. While lenders deposit money in banks in expectation for interests as financial returns, borrowers commit to paying interests on the loans taken from banks. The borrowing interest rate is usually higher than the lending interest rate. Banks make profits from the spread between the borrowing and lending interest rates. The money market refers to financial activities of short-term lending and borrowing, generally for a period of less than a year. Common money market instruments include treasury bills, certificate of deposits, etc.

As the DeFi industry is still at its early stage, most lending/borrowing activities are counterparts of money market instruments in the traditional financial industry. Banking services continue to expand everyday on blockchains. Just like cryptocurrency exchanges, there are also two types of cryptocurrency lending—centralized and decentralized.

Centralized lending firms, such as BlockFi, Celsius, and Voyager Digital, operate in a similar way to banks in the traditional financial industry, which explains why they are also called crypto banks. They take custody of cryptocurrencies from clients and offer loans often to institutional investors. Centralized lending enjoys the same benefits as centralized exchanges such as professional services and high efficiency achieved through the economy of scale. However, customers may lose their principals if the lending firms are hacked or make bad loans. When crypto prices tanked in April 2022, there were "bank runs" on many lending firms, which caused them to freeze customers' withdrawals and to file bankruptcy to restructure their debts. The business model of centralized lending is arguably contradictory to the fundamental philosophy of DeFi, which is to remove the intermediaries.

Decentralized lending protocols, such as Aave and Compound, allow customers to lend and/or borrow in a permissionless and non-custodial way. They do not require identity authorization such as KYC and AML. Lending protocols are accessible at any time and anywhere in the world with Internet connection to blockchains. Because clients take custody of their own cryptocurrencies offered to lending protocols, they can withdraw at any time any amount at their discretion. Decentralized lending protocols also tend to charge lower fees as all transactions are automated without intermediaries. The process of earning interests by lending cryptocurrencies is also called yield farming.

Take Compound as an example. On the lending side, customers convert their cryptocurrencies to cTokens one-to-one. If one deposits one DAI, the person receives one cDAI. If one offers one ETH, then the person receives one cETH. The cTokens represent customers' balance in the protocol and accrue interests over time. Interest is not directly distributed to customers but rather accrues on cTokens which are redeemable at any time. Say, if one cDAI is worth 1.1 DAI after a month, then the customer can redeem DAI to receive a 10% return on the initial investment. On the borrowing side, customers have to supply cryptocurrencies to the protocol as collateral for loans. Each cryptocurrency requires a different collateral factor. For example, if the borrow collateral factor for ETH is 85% and a customer supplies a $100 worth of ETH as collateral to borrow DAI, then he or she can borrow up to 85 DAI. If the collateral value goes below the required collateral factor, which can happen if the ETH price drops drastically in a very short period, then the collateral will be partially sold off by a smart contract along with some liquidation fee. The purpose of overcollateralization is to protect lending protocols from taking on counterparty risk.

There are several reasons why an investor may put one cryptocurrency as a collateral and borrow another cryptocurrency on lending protocols rather than simply trading them on an exchange. First, an investor may not want to give up the ownership of the collateralized cryptocurrency while meeting the needs to use another. This is the same reason as people put their properties as collateral for taking out a loan in the traditional financial industry. People may also use the borrowed cryptocurrencies for speculative investment. Second, the trade of cryptocurrencies requires the sale of one currency, which is then treated as capital gain for tax purposes. Investors may use lending to avoid or delay taxation.

Autonomous Interest Rate

As a decentralized protocol, all lending and borrowing activities are expected to be automated. Hence interest rates are also set algorithmically by a smart contract. Each protocol uses its own version of interest rate setting mechanism. We use Compound as an example to illustrate such an automated process.

Interest rates on Compound are determined by a function of a metric, known as the utilization rate, which is defined as

$$\text{Utilization Rate} = \frac{\text{Amount Borrowed}}{\text{Total Supply}}.$$

The mechanism is motivated by an economic principle, the law of demand and supply. When the demand for a commodity increases under a fixed supply, it drives up the price of the commodity. In the case of capital, the price is measured by interest rate. A high utilization ratio indicates a strong demand. In order to encourage more supply and to suppress additional demand, the protocol has to offer higher lending and borrowing interest rates. This logic explains the positive linear relationship between interest rates and utilization rates in the formulas below. For example, if the size of the total supply of USDC in the protocol is $1,000$ and one wants to borrow 100 USDC, then the utilization rate is $100/1000 = 10\%$. The borrowing interest rate is given by

$$\text{Borrow Rate} = \text{Multiplier} \times \text{Utilization Rate} + \text{Base Rate}.$$

For example, the base rate is 2.5% and the multiplier is 20% in the 2019 version of the Compound whitepaper. Hence, the borrowing interest rate for the example above is $20\% \times 10\% + 2.5\% = 4.5\%$. The lending interest rate is implicitly determined through the exchange rate of cTokens

$$\text{Exchange Rate} = \frac{\text{Available Balance} + \text{Total Borrow Balance} - \text{Reserve}}{\text{cToken Supply}}.$$

Reserve refers to the amount the protocol keeps for profit. Continuing with the earlier example, we can calculate the implied lending interest rate. Assume that the available USDC is supplied through one-to-one exchange to cUSDC and that the reserve rate is 10%. There is a total of $1,000$ cUSDC available in the system. Since the borrower takes 100 USDC and is expected to pay 4.5% interest, then the total borrow balance is 104.5 USDC. The reserve is charged to the borrowing interest at $4.5 \times 10\% = 0.45$. The exchange rate of USDC for each cUSDC can be obtained by

$$\frac{900 + 100 \times (1 + 4.5\%) - 100 \times 4.5\% \times 10\%}{1000} = 1.00405.$$

In other words, the lending interest rate is 0.405%. In general terms, the lending interest rate is implicitly determined by

$$\text{Lend Rate} = \text{Borrow Rate} \times \text{Utilization Rate} \times (1 - \text{Reserve Rate}).$$

Going back to the example, the lending interest rate can be written as

$$4.5\% \times 10\% \times (1 - 10\%) = 0.405\%.$$

To check if the borrowing and lending rates are set up properly, we observe that the total interest earned from borrowers is $4.5\% \times 100 = 4.5$ and the total interest paid to lenders is $0.405\% \times 1000 = 4.05$, with the difference being the 10% profit for the protocol, which is used to increase the value of its governance tokens.

9.2.4 Other Protocols

There are three similar terms used for DeFi investment, namely *yield farming*, *liquidity mining*, and *liquid staking*. The first two terms could also be used on other protocols beyond exchanges and lending/borrowing that reward investors for providing liquidity to their platforms. Therefore, the two terms are sometimes used interchangeably.

Liquid staking becomes popular with the increasing popularity of proof-of-stake mechanism in blockchains. In contrast with liquidity mining where cryptocurrencies are used to enhance the liquidity for protocols with certain financial utility, liquid staking refers specially to the use of cryptocurrencies for the purpose of staking on proof-of-stake blockchains. Liquid staking is used in contrast with self-staking and exchange staking. *Self-staking* refers to the practice of running a validator on a blockchain, for which one has to stake their cryptocurrencies into a smart contract and cannot unstake them until after some lock-up period. *Exchange staking* refers to the practice of staking through a centralized exchange, which consolidates cryptocurrencies from many users to run validators. Exchanges typically stake a portion of their deposits in order to let users to withdraw cryptocurrencies. Therefore, the yield rate for exchange staking is lower than self-staking. Liquid staking protocols are also pooled staking where users' cryptocurrencies are aggregated and used for staking. The difference is that for liquid staking protocols users receive liquid tokens in exchange for their staked cryptocurrencies. The tokens can be used for trading, lending, or collaterals for other protocols. For example, Lido DAO is one of the largest liquid staking protocol. An investor can deposit ETH into Lido's smart contracts and receive stETH tokens in return representing the investor's staked ETH balance. The staked ETH is pooled together with other investors' stake to run validator nodes and accrues staking rewards. However, unlike self-staking, the stETH tokens can be used for additional investment, as the investor can trade stETH for other cryptocurrencies on an exchange and use them for liquidity mining on other protocols.

9.2.5 DeFi Risks

As with any disruptive technology in its infancy and over a period of rapid growth, the DeFi ecosystem is rifle with scams and frauds. Many argue that financial regulators may use the exposure of frauds and the recent collapses of crypto exchanges to crack down on crypto activities and tighten regulations on DeFi services. Most of the regulations on DeFi have so far been limited to anti-money-laundry compliance, crypto fundraising through security laws. However, the distributed ledger technology enables peers to transact with each other through smart contracts without going through intermediaries under the scrutiny of regulators. The DeFi market cannot be truly regulated in the same ways as traditional financial markets. Even if the DeFi market can be regulated as traditional finance, some risks are unavoidable and market participants should understand the risks involved.

We shall adopt the risk taxonomy developed by Chang et al. (2022) here and provide a brief overview of DeFi risks.

- **Smart contract technical risk**: Like any computer software, a smart contract consists of tens of thousands of command lines. There could be bugs and logical defects in the codes causing a smart contract to operate in unexpected ways.
- **Smart contract economic design risk**: There could be intentional or unintentional designs by developers of economic incentives mechanism that can harm the interests of certain protocol participants. A well-known example is a type of crypto scam known as a rug pull, in which investors are lured into a seemingly lucrative new project, and collaterals locked in the smart contracts are then sent to developers once the project reaches a certain value. Because the smart contract is pre-programmed and cannot be altered, the traps are pre-meditated and written into the codes.
- **Cybersecurity operational risk**: Even though the execution of smart contracts is automated on blockchain, its maintenance and interactions with clients often involve third parties and human decision-making. This leaves room for cybersecurity risks. For example, hackers may exploit vulnerabilities in communication channels between different protocols and lure victims to authorize transfer of digital assets to the hackers' wallets. Such risks are in general similar to those in web 2 software systems.
- **Blockchain infrastructure risk**: Blockchain is a computer network which uses economic incentives and consensus mechanisms for miners or validators to keep the system functional. The system can break down for many reasons. For example, it can be overwhelmed when transactions are validated at a lower speed than the requests arrive. As of January 2023, the blockchain Solana has suffered partial or full outage at least 7 times since its launch in 2020.
- **Financial risk**: As with any financial services, investors on DeFi protocols are expected to face investment risks such as price fluctuations on their assets due to appreciation, depreciation, inflation, deflation, and other macroeconomic factors. They also have to cope with credit risks and liquidity risks, as some operations can go bankrupt over a short period of time or halt transactions for an extended period, leaving investors unable to cut losses in a violent market.
- **Social and people risk**: DeFi protocols are always created and maintained by a community of programmers and their followers. Participants may suffer losses when people's opinions change about the performance of the protocol or due to the departure of key personnel.
- **Societal risk**: DeFi services provided by centralized entities may be negatively affected by government regulations and politics. Take the initial coin offerings (ICOs) as an example. First appeared in 2013, ICOs were common ways for DeFi companies to raise capital by selling tokens to retail investors. The US regulator, Securities and Exchange Commission, determined in a 2017 investigative report that tokens are securities and should be subject to federal securities laws. The SEC

has since vigorously pursued many companies for unregistered security offerings. ICO activities have largely cooled down in the US market.

The emergence of DeFi risks gives rise to DeFi insurance. Most of insurance products in the current market are largely provided to mitigate certain smart contract technical risks, economic design risks, and operation risks. We shall discuss the practical details of DeFi insurance in the next section.

9.3 DeFi Insurance

DeFi insurance is a collection of protocols that provide insurance coverage using blockchain technology. It is sometimes referred to as *decentralized insurance* (DeIn). As we have shown throughout this book, there are other decentralized forms of insurance that have been developed in practice without blockchain technology. Therefore, we shall save the term decentralized insurance for the broader umbrella of both blockchain and non-blockchain business models that aim to reduce the role of intermediaries and to improve efficiency in the insurance industry.

Most insurers in the DeFi space provide coverage against smart contract risks. The top insurers include Armor, Nexus Mutual, Unslashed, Sherlock, InsurAce, Risk Harbor, and Bridge Mutual. Armor is a decentralized brokerage that provides coverage underwritten by the Nexus Mutual protocol. As of August 2022, Armor and Nexus Mutual have a combined TVL of around 450 million dollars, which is about 80% of the total TVL in the DeFi insurance sector. Ethereum is the only blockchain with about 20 insurance protocols whereas all other blockchains have fewer than a handful.

In this section, we use Nexus Mutual as a primary example to show how the analogs of various components in the traditional insurance ecosystem have been structured for DeFi insurance. We sometimes supplement the discussion on Nexus with other insurance protocols to show alternative solutions.

Nexus Mutual was founded in 2019 and legally structured as a discretionary mutual. It is argued that Nexus is not an insurer as claim payments are not guaranteed but approved at the discretion of the mutual—a community of users. Whenever possible, insurance terminology is avoided in the Nexus Mutual documentation. For example, instead of calling it an insurance policy, they use the term cover. A policyholder is referred to as a covered member, whereas the premium of a coverage is called cover price. Nevertheless, for the ease of presentation, we shall use these terms interchangeably in this book to compare DeFi insurance with traditional insurance. Nexus provides three types of covers, including protocol cover, yield token cover, and custody cover. Protocol cover provides protection for crypto assets locked in protocols. For example, if you deposit ETH in Maker DAO in exchange for DAI and the protocol is hacked, your fund can be protected by the protocol cover. Yield token cover protects against the de-pegging of yield-bearing tokens from the staked tokens. If one obtains cDAI by depositing DAI in Compound and the cDAI de-pegs

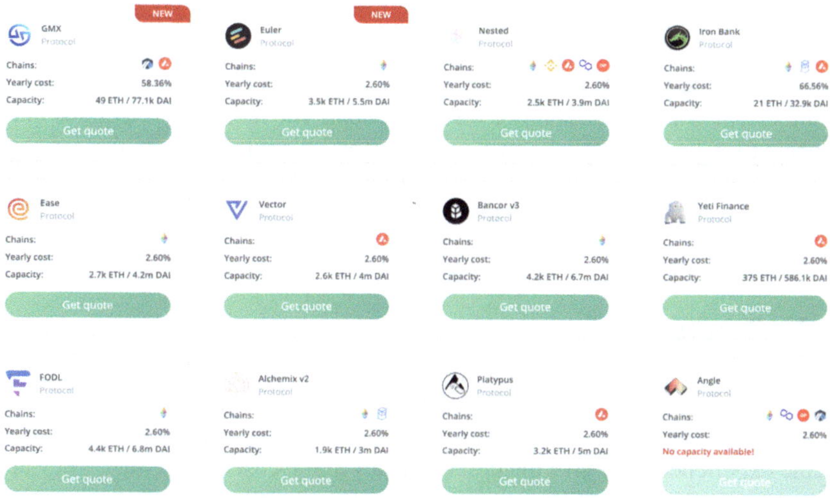

Fig. 9.9 A partial list of protocol covers on Nexus Mutual

in value for more than 10%, then the tokenholder can claim up to 90% of lost value. Custody cover protects users who provide liquidity to protocols that require custody staking. Users can file claims if losing more than 10% of asset due to a hack to the custodian or the custodian halts withdrawals for more than 90 days.

9.3.1 *Underwriting*

In traditional insurance, underwriting is a critical component to measure and assess potential customers' risk profiles and safeguard the insurance pool from excessive exposures beyond its risk-bearing capacity. For example, in health insurance, an underwriter examines an applicant's health records and decides whether the applicant is considered a standard risk and a policy can be issued on standard terms. If the applicant is considered sub-standard, the underwriter specifies exemptions and determines what premium to charge in addition to standard rates.

Because most applications of DeFi insurance are limited to smart contract risks, the underwriting of new covers is either automated or determined through governance procedures. The underwriting is only limited by the pool's capacity. Take Nexus Mutual as an example. There are about one hundred protocols for which one can purchase covers. See Fig. 9.9 for a list of some protocol covers on Nexus. For a listed protocol cover, one can simply purchase a cover by connecting to a crypto wallet and purchasing NXM tokens to buy a cover on the Nexus Mutual website. NXM token is the official governance token for the Nexus Mutual protocol. To list a new protocol cover, a member makes a written proposal on Nexus Mutual's online forum.

Any proposal coordinated with the Nexus core team and its advisory board can be put forth for an on-chain voting. Once members' voting approves the proposal, a new cover can be provided. The premium rate is determined through a decentralized pricing mechanism, which shall be discussed next. The mechanism requires mutual members to stake NXM tokens on protocols to be covered. The mutual places limits on the amount of cover on each specific protocol. The current policy is the total amount of cover cannot exceed 4 times the amount of staking on the protocol.

9.3.2 *Pricing*

Pricing in the traditional insurance industry is always based on centralized models. Actuaries employ statistical tools on historic industry-wide data and company-specific experiences and give the best estimates of the cost of coverage. They apply some pricing rules and build certain margins of error and profit into the premiums. Such an approach is not feasible for smart contract risks, as the DeFi industry is merely a few years old and claims data are scarce.

DeFi insurance often uses decentralized alternatives to the traditional centralized pricing. The cost of coverage is determined by the amount of staked tokens against the underlying assets to be protected. A stake can be viewed as a collateral for a bet on the safety of the underlying assets. Hence, we shall refer to these rules as *pricing by stake*.

Role of risk assessors

As DeFi insurance reduces the role of traditional insurance professionals such as actuaries or underwriters, these key functions are distributed among community members with some knowledge of embedded risks, who are called *risk assessors*. They are economically incentivized to assess the riskiness of insurance products. Risk assessors are required to stake tokens against products in any amount of their choosing. For example, a user can join the Nexus community by purchasing NXM tokens. The NXM tokens can then be used for staking. Rewards and losses for risk assessors are proportional to the stake placed. A high stake represents a risk assessor's confidence on the profitability of the product. Vice versa a low stake indicates the limited trust on the safety of the underlying assets. The final price of the product is often determined by some decreasing function of the collective staked amount from all risk assessors. In other words, the total stake reflects the community's perception of the risk and a low price is offered when high stake is locked in the smart contract. The mechanism of pricing is driven by the token supply of risk assessors, which is similar to that of interest rate setting for lending protocols in Sect. 9.2.3.

Reward

After a cover is sold, a percentage of the premium, 50% for Nexus Mutual, is used as rewards for risk assessors. There are two competing factors in the thought process for staking. The distribution of a premium among risk assessors is done in proportion to their stakes. If a risk assessor perceives the risk as being low, she may choose to stake a high amount to receive as much premium as possible. However, a high stake often leads to a lower unit price by some pricing formula, eroding the profitability of the cover. Each participant has to weigh the benefit and the cost of staking and compete with each other to maximize their profits.

Loss

If a claim is approved, the stake of risk assessors who staked against the cover is burned on a proportional basis up to the claim amount and hence effectively providing the reserve to pay for claims. If the stake is not enough to pay for all claims, all stakes must be burned and any remaining amount shall be absorbed by the mutual as a whole.

In addition to the potential loss of stake due to claims, the locked stake is also subject to market risk. As token prices fluctuate, risk assessors take some downside risks during the lock-in period.

Pricing by stake

While all DeFi insurance platforms use the general principle that policies with high staking are considered low risk, they differ greatly in the actual pricing formulas. In most cases, the actual premium is linear in the cover amount and the period of coverage with respect to the risk cost. In other words, the risk cost is the price of coverage for a unit cover amount and a time unit (typically a year). We use Nexus Mutual and Bridge Mutual as examples of pricing by stake.

- *Nexus Mutual*: A benchmark, called the low-risk cost limit, is used to indicate adequate capital for coverage. The low-risk cost limit is the amount of tokens staked to reach the lowest risk cost. The maximum risk cost is 100%, which means full collateralization (one dollar premium for one dollar cover amount). The lowest risk cost is 2%. The actual risk cost is determined by the following formula and bounded by the lowest and the highest risk costs.

$$\text{risk cost} = 1 - \left(\frac{\text{amount staked}}{\text{low-risk cost limit}}\right)^{1/7}.$$

  Risk assessors receive 50% of all premiums in exchange for their staking on a proportional basis.
- *Bridge Mutual*: For each type of coverage, Bridge Mutual sets up a coverage pool, in which investors stake stablecoins such as USDT as initial capitals. The more capitals available the more capacity for selling coverage. The risk cost is determined by the availability of capital in the coverage pool. When the available capital is in shortage, the risk cost for new coverage is expected to be high. The platform uses the utilization ratio as a measure of capital adequacy.

$$\text{utilization ratio} = \frac{\text{amount locked for existing and new policies}}{\text{total amount staked in coverage pool}}.$$

  If the utilization ratio is below a high threshold (80%), the capital is considered adequate and the risk cost is proportional to a low utilization maximum cost (10%).

$$\text{risk cost} = \text{low utilization max cost} \times \frac{\text{utilization ratio}}{\text{utilization ratio threshold}}.$$

If the utilization ratio is above the threshold, the risk cost is determined linearly between a low utilization cost (10%) and a high utilization maximum cost (50%), i.e.

$$\begin{aligned}\text{risk cost} =& \text{low utilization max cost} + \frac{\text{utilization ratio} - \text{utilization ratio threshold}}{100\% - \text{utilization ratio threshold}} \\ & \times (\text{high utilization max cost} - \text{low utilization max cost}).\end{aligned}$$

Risk assessors are entitled to 80% of all premiums paid. Each assessor receives tokens in proportion to the amount of his/her stake in the coverage pool.
Observe that the use of utilization ratio for pricing has the same logic as that for interest rate setting in the Compound lending protocol.

A natural question is whether such a pricing mechanism truly reflects the risk embedded in the underlying assets. An answer to this question has been provided in Feng et al. (2022a). They observe that, while the protocols have different ways of pricing by stake, they all regard the risk cost (unit size premium) as a decreasing function of the amount staked by risk assessors. As the pricing depends on individuals' staking strategies, the behaviors of all risk assessors are modeled by a Nash equilibrium. They find that the equilibrium price is determined by the most optimistic estimate of loss probability. Without sufficient claims data, the market is dominated by optimists. In order to prevent the price from falling below zero, a floor has to be imposed. However, such prices are often departures from the true underlying risks. This result is consistent with the observation from market data that the majority of covers are offered at the lowest possible price. It is yet to be seen if such a pricing mechanism is sustainable over the long term.

9.3.3 *Capital Management*

Like traditional insurers, DeFi insurance protocols also try to set aside capital to absorb excessive losses. Nexus Mutual uses the concept of Solvency II capital to set its Minimum Capital Requirement ("MCR"), which is the minimum amount of funds the mutual needs to hold to underwrite existing covers.

According to Nexus Mutual's whitepaper, the proposed model of MCR consists of two components. The best estimate liability refers to the expected loss on each individual cover, whereas the buffer represents the additional capital available to absorb excessive losses in extreme events. In other words, the best estimate liability is the analogy of reserve in traditional insurance and the buffer is the equivalent of additional solvency capital.

$$\text{Min Cap Requirement} = \text{Best Liab Estimate} + \text{Buffer}.$$

The whitepaper states that, although the MCR is intended to safeguard the mutual from 1-in-200-year adverse event based on the Solvency II methodology, the mutual uses some deviation from the standard approach on the MCR.

$$\text{Buffer} = \text{Smart Contract Cover Module} + \text{FX Module}.$$

The Smart Contract Cover Module is based on the exposures on the covers that the mutual has underwritten.

$$\text{SC} \cdot \sqrt{\sum_{i,j} r_{ij} \cdot \text{CA}_i \cdot \text{CA}_j},$$

where CA_i is the total cover amount for each individual protocol and custodian, r_{ij} is the correlation efficient between each pair of contracts, and SC is a scaling factor calibrated to make the capital result as a whole more comparable to a full Solvency II calculation. The correlations between each pair of two contracts are established by parsing the respective verified smart contract code, removing comments and spacing, and establishing the proportion of identical text. It is clear that the smart contract cover module is based on the variance-covariance risk aggregation approach (4.2) in Sect. 4.3.2 for traditional insurance. The FX Module refers to extra capital available to absorb potential losses due to the fluctuation of exchange rates in staked cryptocurrencies.

Since the aggregate capital is difficult to compute without sufficient historic data, the mutual sets up the capital using a simpler capital model

$$\text{MCR} = \max\left\{\text{MCR Floor}, \frac{\text{Cover Amount}}{\text{Gearing Factor}}\right\},$$

where currently the gearing factor is set at 4.8 and the MCR floor stands at $162,424.73$ ETH. The term MCR floor is in fact a "ceiling" to limit the holding of capital. Since the MCR has been capped at the floor since November 2022, there is less than 21% of MCR available for the covers underwritten by the mutual. This approach is in fact a formula-based solvency capital requirement similar to those discussed in Sect. 4.2.3 for traditional insurance.

According to Nexus Mutual's whitepaper, "the MCR is calibrated to achieve a 99.5% probability of solvency over a one year period". However, the mutual does not currently implement the Solvency-II-like minimum capital calculation and there have not been long enough records to justify the adequacy of less than 20% solvency capital. Such a statement lacks statistical evidence.

The mutual does hold additional capital in excess of the minimal capital requirement. It uses the coverage ratio to measure the adequacy of funding, defined by

$$\text{MCR\%} = \frac{\text{Size of Capital Pool}}{\text{Min Cap Requirement}}.$$

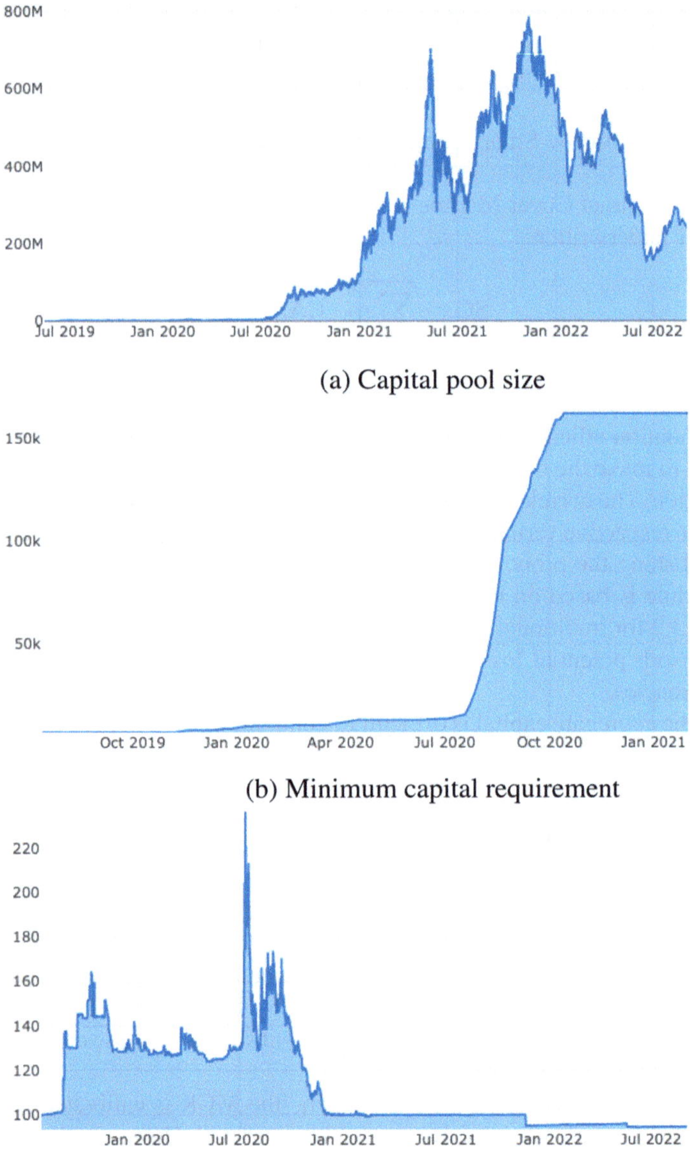

(a) Capital pool size

(b) Minimum capital requirement

(c) Coverage ratio (in percentage)

Fig. 9.10 Nexus capital adequacy. *Source* https://nexustracker.io

The coverage ratio is used for various purposes, including the token price to be discussed in the next section. Figure 9.10 shows the capital pool size, minimum capital requirement, and coverage ratio of Nexus mutual to date. In theory, the capital pool should not be less than the MCR and the coverage ratio should never be less than 100%. However, the current coverage ratio sits at around 94%. It is questionable if the minimum capital requirement mechanism really serves its purpose.

9.3.4 *Asset Management*

Asset liability management is a critical component of the traditional insurance business. Premiums collected by an insurer from policyholders do not necessarily need to be used to pay claims right away. As claims trickle in from period to period, an insurer makes a scheme of payments and structures its reserves accordingly. Reserves are backed by liquid assets as they are expected to be paid out to meet insurance obligations. Additional capitals are then invested in yield-earning assets.

Smart contract covers range from several days to a few months. Due to the relatively short duration of liabilities, most DeFi insurance protocols use a relatively simple asset management strategy, which is through a buy-and-hold liquidity mining with Uniswap. For each liquidity pool to which the mutual provides crypto assets, the investment is limited to a pre-specified range, which is set through a governance mechanism.

9.3.5 *Claims*

When claims are filed in traditional insurance, insurance companies send claims adjusters to inspect the extent of damage and to assess the compensation. Adjusters act in the interest of insurance companies. If there is a dispute between the insurer and the policyholder, then an appraiser is appointed by the insurer and the policyholder to ascertain and state the true value of the damage. The appraiser effectively provides a second evaluation after the adjuster has made an initial determination. When an insurer suspects a fraudulent or criminal activity, it may hire an insurance investigator to conduct research, do surveillance work, and examine evidence of suspicious activities.

DeFi insurance offers similar procedures in a decentralized way. When a policyholder purchases a cover, 10% of the premium is locked as a deposit for claims. If a claim is approved, then the deposit is returned to the policyholder. If a claim is denied, then the deposit is burned. The deposit is used as a deterrent to fraudulent claim. When claims are submitted, they are handled by a community of claims assessors through a consensus mechanism. A policyholder can submit a second claim, which shall be voted by a different set of claims assessors.

Like risk assessors, claims assessors are also members with the knowledge of smart contract claims. They have to stake their NXM tokens in order to participate

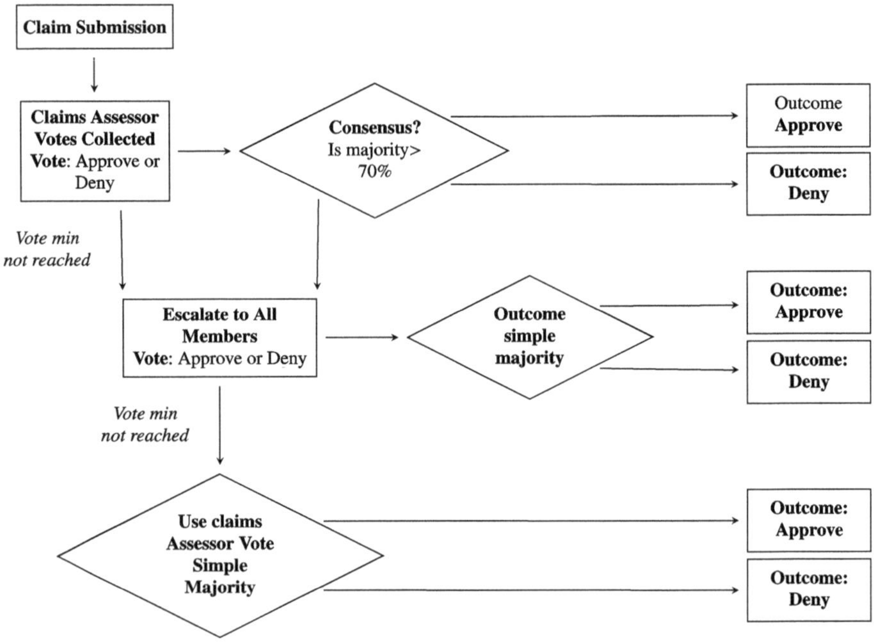

Fig. 9.11 Claims assessment process

in the claims assessment process as shown in Fig. 9.11. Their votes are weighted by their stake amount. If the super-majority (70%) votes to approve or deny a claim, then the claim is settled. Otherwise, the voting is escalated to all members. If a quorum is secured, the result is determined by a simple majority. Otherwise, the voting returns to claims assessors for a simple majority (50%) decision. To encourage claims assessors to do due intelligence, those who voted with the majority shall be rewarded by 20% of the premium, each assessor in proportion to their stake. It should be pointed out that if there is no claim, then the initial deposit for claim submission is returned to the policyholder and the 40% of premium is sent to the capital pool. If an assessor votes against the majority, he is not penalized directly as it could be due to a genuine difference in opinions. However, the assessor's stake is locked for 7 days in the Nexus model. Any locked stake is subject to market risk, as the token price may be volatile.

Many other insurance protocols use similar claims assessment mechanisms. There are also some minor differences. Bridge Mutual for instance has a reputation score for each claims assessor. Each assessor's voting power is determined by their stake multiplied by their reputation score. Hence it is in the interest of the claims assessor to increase their reputation score. If an assessor votes with the majority, she gains a reputation score, which is linear with respect to the percentage of majority votes. If an assessor votes with the minority, he loses a reputation score, which is quadratic in its distance from 50%. The smaller the minority percentage, the bigger penalty on the reputation score. A claimant can also request a second evaluation in an appeal process, in which only trusted voters, top 15% of risk assessors by reputation score, can participate.

The first cover on Nexus Mutual started on July 12, 2019. As of September 15, 2022, there have been a total of 7, 774 covers sold through the Nexus Mutual protocol and 121 claim submissions, of which 30 have been approved. While the statistics indicate a very low probability of a successful claim, the claims history is not long enough by the standards of the traditional insurance industry to provide a credible statistical evidence. It is yet to be seen if the capital model is sufficient to safeguard the protocol against "black swan" events as claimed.

9.3.6 Governance

Many DeFi insurance protocols including Nexus Mutual are legally structured as a decentralized autonomous organization (DAO), where tokenholders participate in its management and decision-making. In theory, there is no central authority in a DAO and the shared governance is carried out through tokenholders. DAO typically has a shared treasury and a governance structure with rules ratified and enforced on-chain. However, in practice, it is difficult to have every member to vote on every decision. Therefore, each DAO has a somewhat different process for decision-making. It is common to categorize decisions by their importance to the functioning of a DAO and to assign the right stakeholder to make decisions. Decisions are typically token weighted.

In the case of Nexus Mutual, decisions are put into four categories, including (1) day-to-day operations; (2) upgrades, technical changes, and use of funds; (3) emergency actions; and (4) critical decisions. Actions in the first category are all automated and executed by smart contracts. The second category tends to be more technical and requires specialized knowledge. Hence the mutual designates an advisory board to initiate changes but does members the ability to vote on changes. The third category requires time-sensitive decision-making and hence all actions are delegated to the advisory board. The last category involves the most critical decisions, which can be viewed as changes to the "constitution" of the DAO, and hence requires high bar for changes from status-quo, including high quorum and super-majority.

9.3.7 Tokenomics

Tokenomics is a portmanteau of "token" and "economics". Protocols design tokenomics to provide incentives for healthy activities and use penalties to deter harmful activities. It is in some sense similar to equities in stock companies, where holders can participate in the growth in equity and/or voting power for key decisions. However, tokens can be much more than equities. For example, their functions can be divided into several categories, including utility tokens, security tokens, payment tokens, non-fungible tokens, etc. Utility token refers to the type of tokens that are used for access to certain services. For example, NXM tokens can be used for risk assessment, claims assessment, deposit for claims assessment, and governance. Security token represents a share in equity, a stake, and voting rights on the underlying asset. In this sense, NXM is also a security token, as it can represent a member's ownership of the protocol. Payment tokens are used for buying goods and services. For example, ETH and Bitcoin are primary examples of payment tokens. A non-fungible token is a certificate of ownership to a unique asset on a blockchain.

For the Nexus Mutual protocol, NXM tokens are given prices according to a pre-specified formula

$$\text{Token Price} = A + \frac{\text{MCR}}{C} \cdot \text{MCR\%}^4,$$

where A and C are constants used to caliber the token price to an acceptable price at the time of issuance. The formula was structured in such a way to incentivize the "auto-correction" of minimum capital requirement. When claims are paid, the capital pool shrinks in size more so than the MCR and hence lowers the coverage ratio (MCR%). The formula suggests that the token price shall go down. The low token price should encourage more investors to buy tokens at a discount and new cash is expected to be injected into the capital pool. When covers are sold, the capital pool increases faster than the MCR, pushing up the coverage ratio (MCR%) and resulting in exponential growth in token price. A high token price is expected to discourage investors from buying the tokens, which avoids excessive capitals.

However, it should be pointed out that the tokenconomics does not work anymore for Nexus Mutual. As one can observe from Fig. 9.10, the coverage ratio (MCR%) has already reached below 100% and the MCR is capped at a fixed level. Hence the token price by its design cannot drop further. The mechanism has lost its function in design to adjust the demand and supply of tokens to maintain some level of stability for its capital pool.

While the pricing formula of NXM sets the token price at around \$50, there is close to none trading volume. NXM tokens can only be traded among members by design. The wNXM (wrapped NXM) is used to trade outside the mutual and its price is driven freely by supply and demand. The wNXM can be exchanged one-to-one with NXM. Almost all new NXM tokens are bought today through exchanges between wNXM (wrapped NXM) tokens and NXM tokens. The current price of wNXM is at around \$16. The price gap between the NXM and the wNXM offers little incentives for new inflow of capital, which partly explains the stagnant capital pool size as shown in Fig. 9.10.

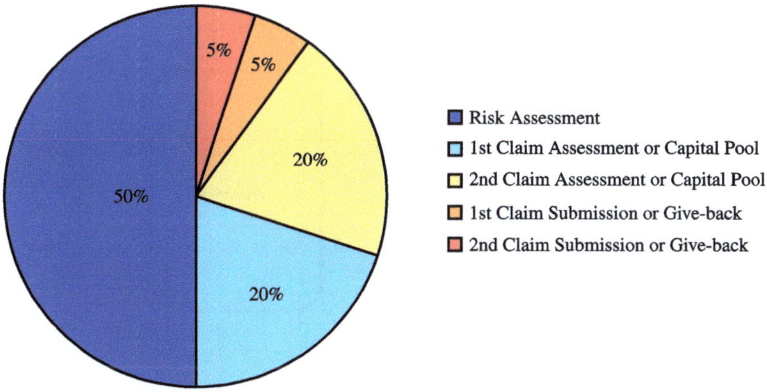

Fig. 9.12 Breakdown of premium

9.3.8 Risk Sharing

Although the mechanism of risk sharing is never explicitly stated in the documentation for DeFi insurance protocols, we can infer from other components how risks are spread out among various stakeholders. We take a minimalist approach in this section to identify only the core elements in the relationship for risk sharing. There are typically four stakeholders in the aforementioned insurance protocols, namely risk assessor, claims assessor, policyholder, and tokenholder through the protocol itself.

As in any economic incentives mechanism, every risk-bearing activity is encouraged with a reward. Let us summarize the breakdown of premium as rewards for various stakeholders. Figure 9.12 shows the percentage of premium allocated for different purposes for a Nexus mutual cover. As alluded to earlier, a half of the premium goes to risk assessors, and the other half is split among other stakeholders. Each policyholder can file a claim twice. For each claim, 20% of the premium is allocated to claims assessors in proportion to their stake. For each successful claim, 5% of the premium is returned to the policyholder. If a claim is unsuccessful, the 5% is confiscated by the capital pool. If there is no claim filed, the 10% returns to the policyholder and the 40% goes to the capital pool.

As one can tell from the capital management component, most DeFi insurance protocols act in many ways like traditional insurers. They set up capitals to absorb tail risks in a similar way to solvency capital. However, the underlying risks are managed in similar ways to decentralized insurance models discussed in previous chapters. Figure 9.13 offers an overview of risk management techniques embedded in the Nexus mutual protocol. After policyholders purchase covers/policies, the risks are effectively transferred from policyholders to the protocol and aggregated into a coverage pool. Risk assessors take a similar role to a primary insurers in the traditional insurance industry. They stake tokens and provide initial capitals to support the

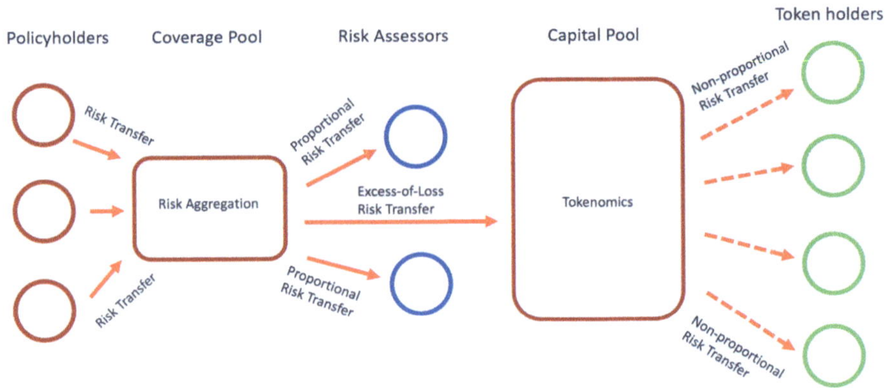

Fig. 9.13 Risk management techniques in Nexus Mutual protocol

covers. If claims are made, their tokens are burned and the locked collateral (in ETH) is used to pay for the claims. In this way, the risks (rewards and losses) are absorbed in proportion to their stakes. The protocol often acts like a reinsurer that offers an excess-of-loss policy for all losses beyond the risk-bearing capacity of risk assessors. The capital pool accumulates all profits from previous periods and uses them to pay for losses after all risk assessors' stakes have been burned. Most protocols use some tokenomics models to pass on profits and losses to their investors by adjusting the price of tokens. Some are straight pass-through, whereas others use the capital pool as a cushion and algorithmically adjust the token prices to pass on some risks.

Let us formulate the risk sharing rule behind typical DeFi insurance protocols. Like traditional insurance, the risks underwritten in DeFi insurance protocols are also pooled together before the sum is allocated to various stakeholders. In exchange for the coverage, the policyholders pay fixed premiums as described earlier with the pricing-by-stake approach. Denote the risks from all policyholders by $(X_1, \cdots, X_n)$ and the premiums from policyholders, respectively, by $(\pi_1, \cdots, \pi_n)$. The aggregate risk from all policies is given by $S = \sum_{i=1}^{n} X_i$ and the total premium is given by $\pi = \sum_{i=1}^{n} \pi_i$. As illustrated in Fig. 9.13, risk assessors act as the first layer of risk sharing. Should there be any claim, the stakes provided by risk assessors are burned before any excessive amount goes to the protocols. Let us denote the proportion of premium that is awarded to risk assessors by α. Recall from the earlier formulation that risk assessors are assumed to stake the amount $(V_1, \cdots, V_n)$ and the community collectively stake the total amount $V = \sum_{i=1}^{n} V_n$. Therefore, each risk assessor takes on the risk

$$\frac{V_i}{V} S \wedge V_i - \frac{V_i}{V} \alpha\pi = \frac{V_i}{V} (S \wedge V - \alpha\pi).$$

Risk assessor community collectively takes the risk

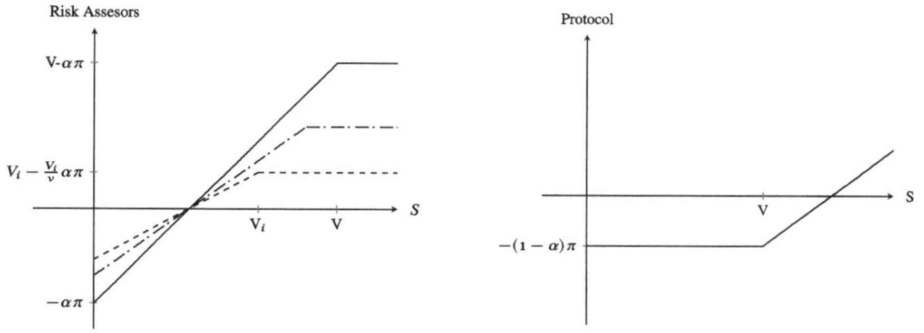

(a) Proportional risk transfer to risk assessors (b) Excess-of-loss risk transfer to capital pool

Fig. 9.14 Risk transfers in DeFi insurance protocol

$$\sum_{i=1}^{n} \frac{V_i}{V}\left(S \wedge V_i - \frac{V_i}{V}\alpha\pi\right) = S \wedge V - \alpha\pi.$$

In practice, the remaining premium may go to claims assessors if a claim is filed or return to policyholders to discourage fraudulent claims. Nevertheless, claims are infrequent and we shall remove the claims assessment cost and potential fee income to the protocol from unsuccessful claims. For simplicity, we only consider that all remaining premiums go to the protocol. Therefore, the aggregate risk borne by the protocol can be formulated by

$$(S - V)_+ - (1 - \alpha)\pi.$$

Figure 9.14 shows the payoffs for risk assessors and that for the protocol. The two types of risk transfer schemes have been discussed in Sects. 3.3.3 and 3.3.4. This DeFi insurance protocol is yet another example to show how common these risk transfer strategies have been used in real-life applications.

The risk sharing does not stop with the protocol. The risks are ultimately carried by all tokenholders of the protocol. There are several ways in which the risk sharing has been extended to tokenholders. The simplest approach is a pass-through mechanism to tokenholder in which the token price is linked to the capital pool size. In other words, if the capital pool size increases by 10%, so does the token value. In such a case, the aggregate risk undertaken by the protocol is then proportionally allocated to all tokenholders. Another approach is to indirectly pass on gains or losses to tokenholders. In the case of Nexus Mutual, the capital pool preserves gains and absorbs some losses from previous periods and acts effectively as a cushion to smooth out losses and gains from policies. This logic is very similar to those of takaful models discussed in Sect. 5.4. Since the discussion is limited to one-period models in this book for simplicity, we assume that the balance of the capital pool is given by U. We can think of the minimal capital requirement (MCR) as a function of the underlying risks, say, MCR= $f(X_1, \cdots, X_n)$. Therefore, the coverage ratio can be represented by

$$\text{MCR\%} = \frac{U - (S - V)_+ + (1-\alpha)\pi}{f(X_1, \cdots, X_n)}.$$

The token price indicates how much risk is passed on to tokenholders

$$A + B[U - (S - V)_+ + (1-\alpha)\pi]^5,$$

for some constant B. Note that this is not a direct pass-through approach because the sum of token values need not match the size of the capital pool. It is clear from the formulas that any excessive loss absorbed by the protocol is partially passed on to tokenholders. It is yet to be seen if such a tokenomics model would be economically sustainable in the long run.

A natural question to ask is whether the risk sharing of DeFi insurance fits into the unified framework of decentralized insurance developed in Chap. 8. The answer is yes for a single-period model. The overarching condition for decentralized insurance discussed in earlier chapters is the loss conservation before and after risk exchange among a closed group of n participants,

$$\sum_{i=1}^{n} X_i = \sum_{i=1}^{n} Y_i,$$

where X_i and Y_i represent the pre-exchange risk and the post-exchange risk for the i-th participant. The risk sharing in the context of DeFi insurance is an extension of such a problem where the losses are passed on from a group of policyholders to another group of investors. There are often two types of investors in DeFi insurance protocols—pool-specific underwriting investors and aggregate tail risk investors. They play similar roles as primary insurers and reinsurers in the traditional insurance industry. Pool-specific underwriting investors provide stake for a particular pool (or a particular risk), whereas aggregate tail risk investors (a.k.a tokenholders) put their investment in tokens backing an aggregate capital pool, which can provide liquidity to different risk pools. Formulated in a similar way, we can write the loss conservation condition as

$$\sum_{i=1}^{n} X_i = \sum_{j=1}^{m} Y_j + \sum_{k=0}^{l} Z_k,$$

where Y_j can represent the risk undertaken by the j-th pool-specific investor and Z_k the risk carried by the k-th tokenholder for a group of m risk assessors and another group of l tokenholders. We use Z_0 to denote the residual risk for the capital pool. Take Nexus Mutual for an example. The pre-exchange risks are aggregated and pooled together by the underlying protocol, $S = \sum_{i=1}^{n} X_i$. The post-exchange risk underwritten by each risk assessor is in proportion to his/her stake and given by

$$Y_i = \frac{V_i}{V}(S \wedge V - \alpha\pi), \qquad \text{for } i = 1, \cdots, n.$$

Suppose that the token price at the inception of the coverage for a particular protocol under consideration is P_0. As stated earlier, the new token price is then given by

$$P_1 = A + B[U - (S - V)_+(1 - \alpha)\pi]^5,$$

for some constants A and B. Assume that the number of tokens held by k-th tokenholder is n_k, then the total risk passed on the k-th tokenholder can be written as

$$Z_k = n_k(P_1 - P_0), \qquad \text{for } k = 1, \cdots, l.$$

Any residual risk is held in the capital tool, i.e.

$$Z_0 = (S - V)_+ - (1 - \alpha)\pi - \sum_{k=1}^{m} n_k(P_1 - P_0).$$

The residual risk is in essence carried on by investors in future periods.

In theory, many of the same risk sharing strategies discussed in Chaps. 6 and 7 can also be used to define the risk distribution among investors. However, the innovative approaches taken by practitioners are often driven by other practical considerations than utility or risk measure optimization. There are many other DeFi insurance protocols that structure the risk allocation in different ways. Ensuro, a crypto capital provider, underwrites off-chain risks such as travel cancelation ("cancel for any reason") policies and uses crypto assets for insurance capitals. The losses and profits from a coverage pool are passed on to investors through tokens in a way similar to the aforementioned DeFi insurance protocols. However, in contrast with Nexus Mutual, Ensuro structures its capital pool into two tranches with different priorities for claims. Losses are first applied against the capital in the junior tranche. When the junior tranche runs out of capacity, the excessive losses are moved to the senior tranche. The financial returns on investments in these two tranches are expected to be commensurate with their risks. Again such a loss allocation is in principle consistent with the loss conservation condition described earlier.

One should also bear in mind that most DeFi insurance business models are in fact multi-period models. As stated right from the introduction, we mostly limit our focus to one-period models in this book. Interested readers are referred to a sequel of this book, in which we shall address decentralized insurance and annuities in multi-period models and their industry practices.

More discussions of DeFi insurance and its tokenomics can be found in Cousaert et al. (2022) and Braun et al. (2021).

9.4 Departing Words

The rapid growth of DeFi in the 2020s is often likened to the Gold Rush of the 1840s. People call the DeFi space the "Wild West" of the crypto world. There seem to be unlimited possibilities for anyone in any part of the world to propose new protocols to introduce services that were only possible for established institutional players in the traditional financial industry. As the DeFi industry is merely a few years old, there is little regulation by traditional financial authorities to regulate the financial activities in this space. While there have been tremendous innovations in financial services, the space is also rife with fraud, scams, money laundering, and other criminal activities.

There is a clear need for insurance as a tool of protection for investor's financial interests. Most of DeFi insurance protocols today focus on smart contract risks. However, the blockchain technology can have far-reaching applications beyond the insurance for on-chain crypto assets. There have been ongoing efforts by many startups, such as Etherisk and Ensuro, to introduce DeFi insurance products for off-chain risks, including flight cancelation, crop yields, etc.

The purpose of this book is to offer a discussion of first principles in risk analysis of insurance products. While we can only offer a glimpse of a wide variety of decentralized insurance, the fundamental principles can be applied as a guide for future product design and innovations. As discussed in the section on DeFi insurance, there are many questions to be answered even for a top DeFi insurance protocol like Nexus Mutual. Actuaries, risk analysts, and economists can play a critical role of "bounty hunter" in this Wild West to understand the risks hidden from sophisticated economic incentives and capture illogical or ineffective product designs. We hope this book can help researchers and entrepreneurs understand the underpinning theory behind current practices on decentralized insurance and inspire them to propose innovative business models for a better future of the industry.

References

Abdikerimova, S., Boonen, T. J., & Feng, R. (2022). *Multi-period peer-to-peer risk sharing*. Available at SSRN: https://ssrn.com/abstract=4065099.
Abdikerimova, S., & Feng, R. (2022). Peer-to-peer multi-risk insurance and mutual aid. *European Journal of Operational Research, 299*(2), 735–749.
Achman, L., & Chollet, D. (2001). *Insuring the uninsurable: An overview of state high-risk health insurance pools* (Technical report 472). The Common Wealth Fund.
Adams, H., Zinsmeister, N., Salem, M., Keefer, R., & Robinson, D. (2021). *Uniswap v3 Core* (Technical report). Uniswap.
Annuar, H. A., Rosly, S. A., & Abdul Rashid, H. M. (2004). AlWakalah and its impact on the growth and performance of takaful companies: A Malaysian case. *The European Journal of Management and Public Policy, 3*(1), 84–119.
Ant Group Research Institute. (2020). *White Paper on Online Mutual Aid Industry*. Ant Group.
Arbenz, P., Hummel, C., & Mainik, G. (2012). Copula based hierarchical risk aggregation through sample reordering. *Insurance: Mathematics and Economics, 51*(1), 122–133.
Archer, S., Karim, R. A. A., & Nienhaus, V. (2011). *Concepts and regulatory issues*. TakafulIslamic insurance.
Artzner, P., Delbaen, F., Eber, J.-M., & Heath, D. (1999). Coherent measures of risk. *Mathematical Finance, 9*(3), 203–228.
Baione, F., De Angelis, P., & Granito, I. (2021). Capital allocation and RORAC optimization under solvency 2 standard formula. *Annals of Operations Research, 299*(1–2), 747–763.
Barrieu, P., & El Karoui, N. (2005). Inf-convolution of risk measures and optimal risk transfer. *Finance and Stochastics, 9*(2), 269–298.
Beard, R. E., Pentikäinen, T., & Pesonen, E. (1969). *Risk theory. The stochastic basis of insurance*. Wiley.
Billah, M. M. (1998). islamic insurance: its origins and development. *Arab Law Quarterly, 13*(4), 386–422.
Bollmann, A., & Wang, S. S. (2019). *International catastrophe pooling for extreme weather* (Technical report). Society of Actuaries.
Borch, K. (1960). The safety loading of reinsurance premiums. *Scandinavian Actuarial Journal,* (3–4), 163–184.
Borch, K. (1968). General equilibrium in the economics of uncertainty. In *Risk and Uncertainty: Proceedings of a Conference held by the International Economic Association* (pp. 247–264).
Braun, A., Haeusle, N., & Karpischek, S. (2021). Decentralized autonomous risk transfer on the blockchain. In *The Routledge handbook of fintech* (pp. 69–78). Routledge.
Bühlmann, H. (1980/1981). An economic premium principle. *Astin Bulletin, 11*(1), 52–60. ISSN: 0515-0361.

R. Feng, *Decentralized Insurance*, Springer Actuarial,
https://doi.org/10.1007/978-3-031-29559-1

Bühlmann, H., & Jewell, W. S. (1978). *Unicity of fair pareto optimal risk exchanges*, Technical report. California University Berkeley Operations Research Center.
Bühlmann, H., & Jewell, W.S. (1979). Optimal risk exchanges. *ASTIN Bulletin: The Journal of the IAA, 10*(3), 243–262.
Carlier, G., Dana, R.-A., & Galichon, A. (2012). Pareto efficiency for the concave order and multivariate comonotonicity. *Journal of Economic Theory, 147*(1), 207–229.
Chang, T., Ho, J., Tirrell, Z., Weng, G., & You, J. (2022). *A risk classification framework for decentralized finance protocols. Exploring Emerging Risks for Insurers and Reinsurers*, Technical report. Society of Actuaries.
Chankong, V., & Haimes, Y. Y. (2008). *Multiobjective decision making: Theory and methodology.* Courier Dover Publications.
Charpentier, A., Kouakou, L., Löwe, M., Ratz, P., & Vermet, F. (2021). Collaborative insurance sustainability and network structure. arXiv:2107.02764.
Chen, X., Chong, W. F., Feng, R., & Zhang, L. (2021). Pandemic risk management: resources contingency planning and allocation. *Insurance: Mathematics and Economics, 101*, 359–383.
Chen, Z., Feng, R., Wei, L., & Zhao, J. (2023). Cost-effectiveness, fairness and adverse selection in mutual aid. *European Financial Management,* (Forthcoming).
China Insurance Regulatory Commission. (2013). *China Insurance Industry Critical Illness Experience Rates (2006–2010).* Government of the People's Republic of China.
Chong, W. F., Feng, R., Hu, H., & Zhang, L. (2022). Cyber Risk Assessment for Capital Management.
Chong, W. F., Feng, R., & Jin, L. (2021). Holistic principle for risk aggregation and capital allocation. *Annals of Operations Research,* 1–34.
Cousaert, S., Vadgama, N., Xu, J. (2022). Token-based insurance solutions on blockchain. In *Blockchains and the token economy: Theory and practice* (pp. 237–260). Springer.
Cummins, J. D. (1991). Statistical and financial models of insurance pricing and the insurance firm. *The Journal of Risk and Insurance, 58*(2), 261–302.
Denuit, M. (2019). Size-biased transform and conditional mean risk sharing, with application to P2P insurance and tontines. *ASTIN Bulletin: The Journal of the IAA, 49*(3), 591–617.
Denuit, M. (2020). Investing in your own and peers' risks: The simple analytics of P2P insurance. *European Actuarial Journal, 10*(2), 335–359.
Denuit, M., & Dhaene, J. (2012a). Convex order and comonotonic conditional mean risk sharing. *Insurance: Mathematics and Economics, 51*(2), 265–270.
Denuit, M., & Dhaene, J. (2012b). Convex order and comonotonic conditional mean risk sharing. *Insurance: Mathematics and Economics, 51*(2), 265–270.
Denuit, M., Dhaene, J., Goovaerts, M., & Kaas, R. (2006). *Actuarial theory for dependent risks: Measures, orders and models.* Wiley.
Denuit, M., Dhaene, J., & Robert, C. Y. (2022). Risk-sharing rules and their properties, with applications to peer-to-peer insurance. *Journal of Risk and Insurance, 89*(3), 615–667.
Denuit, M., & Robert, C. Y. (2022). Conditional mean risk sharing in the individual model with graphical dependencies. *Annals of Actuarial Science, 16*(1), 183–209.
Denuit, M., & Robert, C. Y. (2021). Risk sharing under the dominant peer-to-peer property and casualty insurance business models. *Risk Management and Insurance Review.*
Dhaene, J., Goovaerts, M. J., Lundin, M., & Vanduffel, S. (2005). Aggregating economic capital. *Belgian Actuarial Bulletin, 5,* 52–56.
Dhaene, J., Kaas, R., & Goovaerts, M. J. (2003). Economic capital allocation derived from risk measures. *North American Actuarial Journal, 7*(2), 44–56
Dhaene, J., Tsanakas, A., Valdez, E. A., & Vanduffel, S. (2012). Optimal capital allocation principles. *Journal of Risk and Insurance, 79*(1), 1–28.
Eisen, R., & Eckles, D. L. (2011). *Insurance economics.* Springer.
Embrechts, P., Liu, H., & Wang, R. (2018). Quantile-based risk sharing. *Operations Research, 66*(4), 936–949.

Farr, I., Mueller, H., Scanlon, M., & Stronkhorst, S. (2008). *Economic capital for life insurance companies*. Society of Actuaries.

Feng, R. (2018). *An introduction to computational risk management of equity-linked insurance*. English (US), Chapman Hall/CRC Financial Mathematics Series, CRC Press.

Feng, R., Li, M., & Bichuch, M.(2022a). Pricing by stake in DeFi insurance.

Feng, R., Liu, C., & Taylor, S. (2023). Peer-to-peer risk sharing with an application to food risk pooling. *Annals of Operations Research, 321*(1–2), 813–842.

Feng, R., Liu, M., & Zhang, N. (2022b). A Unified Theory of Decentralized Insurance. Available at SSRN: https://ssrn.com/abstract=4013729.

Feng, R., & Shimizu, Y. (2016). Applications of central limit theorems for equity-linked insurance. *Insurance: Mathematics and Economics, 69*, 138–148.

Filipoviĉ, D. (2009). Multi-level risk aggregation. *ASTIN Bulletin: The Journal of the IAA, 39*(2), 565–575.

Friedland, J. (2021). *Fundamentals of general insurance actuarial analysis*. (2nd ed.). Society of Actuaries.

Furman, E., Kye, Y., & Su, J. (2021). A reconciliation of the top-down and bottom-up approaches to risk capital allocations: Proportional allocations revisited. *North American Actuarial Journal, 25*(3), 395–416.

Gatteschi, V., Lamberti, F., Demartini, C., Pranteda, C., & Santamarıa, V. (2018). Blockchain and smart contracts for insurance: Is the technology mature enough? *Future Internet, 10*(2), 20.

Gerber, H. U., & Pafum, G. (1998). Utility functions: from risk theory to finance. *North American Actuarial Journal, 2*(3), 74–91.

Gerber, H. U. (1979). *An introduction to mathematical risk theory* (Vol. 8). S. S. Heubner Foundation Monograph Series. University of Pennsylvania, Wharton School, S.S. Huebner Foundation for Insurance Education.

Ginsburg, M. (1940). Roman military clubs and their social functions. *Transactions and Proceedings of the American Philological Association, 71*, 201–213.

Gómez, F., Tang, O., & Tong, Z. (2022). The gradient allocation principle based on the higher moment risk measure. *Journal of Banking & Finance, 143*, 106544.

Guiso, L. (2012). Trust and insurance markets. *Economic Notes by Banca Monte dei Paschi di Siena SpA, 41*, 1–26.

Hans, R., Zuber, H., Rizk, A., Steinmetz, R. (2017). Blockchain and smart contracts: Disruptive technologies for the insurance market. In *Proceedings of the 23rd Americas Conference on Information Systems*.

Ho, F. (2016). *Investing in sukuk: How does it work*? https://www.imoney.my/articles/investing-in-sukuk-how-does-it-work.

Jouini, E., Schachermayer, W., Touzi, N. (2008). Optimal risk sharing for law invariant monetary utility functions. *Mathematical Finance: An International Journal of Mathematics, Statistics and Financial Economics, 18*(2), 269–292.

Lipton, A., & Treccani, A. (2021). *Blockchain and distributed ledgers: Mathematics, technology, and economics*. World Scientific.

Malik, A., Ullah, K., et al. (2019). *Introduction to takaful*. Springer.

Mangasarian, O. L. (1994). *Nonlinear programming*. SIAM.

Manning, W. G., Newhouse, J. P., Duan, N., Keeler, E. B., & Leibowitz, A. (1987). Health insurance and the demand for medical care: evidence from a randomized experiment. *The American Economic Review*, 251–277.

McNeil, A. J., Frey, R., & Embrechts, P. (2015). *Quantitative risk management: Concepts, techniques and tools*. Princeton University Press.

Miettinen, K. (2012). *Nonlinear multiobjective optimization* (Vol. 12). Springer Science & Business Media.

Milevsky, M. A. (2015). *King William's tontine: Why the retirement annuity of the future should resemble its past*. Cambridge University Press.

Morgan, R. (2020). *What are Shariah compliant investments*? https://www.charles-stanley.co.uk/group/cs-live/what-are-shariah-compliant-investments.

Pazdera, J., Schumacher, J. M., & Werker, B. J. M. (2017). The composite iteration algorithm for finding efficient and financially fair risk-sharing rules. *Journal of Mathematical Economics, 72*, 122–133.

Perry, J. S. (2006). *The Roman Collegia: the modern evolution of an ancient concept*. Koninklijke Brill NV.

Pfeifer, D., & Strassburger, D. (2008). Solvency II: stability problems with the SCR aggregation formul. *Scandinavian Actuarial Journal*, (1), 61–77.

Pistilli, T. (2021). Health care sharing ministries - what are these costing-sharing products in the individual health care market? *The Actuary*.

Powers, M. R., & Shubik, M. (1998). On the tradeoff between the law of large numbers and oligopoly in insurance. *Insurance: Mathematics and Economics, 23*(2), 141–156.

Raddato, C. (2016). *The Natalis Antinoi and the collegium of Diana and Antinous in Lanuvium*. https://followinghadrian.com/2016/11/27/the-natalis-antinoi-and-the-collegium-of-diana-and-antinous-in-lanuvium/.

Saeed, M. (2019). *Effect of takaful models on performance of takaful operators*. Available at SSRN 3500383.

Shaked, M., & Shanthikumar, J. G. (1994). *Stochastic orders and their applications*. Academic.

Soualhi, Y. (2016). Surplus distribution in current Takāful operations: a critical Sharıah perspective. In: *Arab Law Quarterly, 30*(3), 224–244.

SwissLife. (2018). *Multinational Pooling—Global Employee Benefits Solutions*. Swiss Life Global Solutions: Tech. rep.

Wahab, A., Abdul, R., Lewis, M. K., Kabir Hassan, M. (2007). Islamic takaful: business models, Shariah concerns, and proposed solutions. *Thunderbird International Business Review, 49*(3), 371–396.

Index

R. Feng, *Decentralized Insurance*, Springer Actuarial,
https://doi.org/10.1007/978-3-031-29559-1

MIX
Papier aus verantwortungsvollen Quellen
Paper from responsible sources
FSC® C105338
FSC
www.fsc.org

If you have any concerns about our products,
you can contact us on
ProductSafety@springernature.com

In case Publisher is established outside the EU,
the EU authorized representative is:
Springer Nature Customer Service Center GmbH
Europaplatz 3, 69115 Heidelberg, Germany

Printed by Libri Plureos GmbH
in Hamburg, Germany